AF768979

Mobile Computing

Edited by Jesus Hamilton Ortiz

Published in London, United Kingdom

IntechOpen

Supporting open minds since 2005

Mobile Computing
http://dx.doi.org/10.5772/intechopen.80095
Edited by Jesus Hamilton Ortiz

Contributors
Erika Hernández-Rubio, Amilcar Meneses-Viveros, Sonia G. Mendoza-Chapa, José Jailton, Tassio Costa Carvalho, Miguel Itallo B. Azevedo, Carlos Coutinho, Eylon Martins Toda, Thembelihle Dlamini, Larysa Globa, Svitlana Sulima, Mariia Skulysh, Oleksandr Stryzhak, Stanislav Dovgyi, Zsolt P. Ori, Naeem Ahmad, Shuchi Sethi, Pavel Loskot, Salman Al-Shehri, Michael J. Hirsch, V.R. Prakash, K. Sakthidasan Sankaran, G. Ramprabu, Jesus Hamilton Ortiz, Fernando Velez Varela, Bazil Taha Ahmed

© The Editor(s) and the Author(s) 2020
The rights of the editor(s) and the author(s) have been asserted in accordance with the Copyright, Designs and Patents Act 1988. All rights to the book as a whole are reserved by INTECHOPEN LIMITED. The book as a whole (compilation) cannot be reproduced, distributed or used for commercial or non-commercial purposes without INTECHOPEN LIMITED's written permission. Enquiries concerning the use of the book should be directed to INTECHOPEN LIMITED rights and permissions department (permissions@intechopen.com).
Violations are liable to prosecution under the governing Copyright Law.

(cc) BY

Individual chapters of this publication are distributed under the terms of the Creative Commons Attribution 3.0 Unported License which permits commercial use, distribution and reproduction of the individual chapters, provided the original author(s) and source publication are appropriately acknowledged. If so indicated, certain images may not be included under the Creative Commons license. In such cases users will need to obtain permission from the license holder to reproduce the material. More details and guidelines concerning content reuse and adaptation can be found at http://www.intechopen.com/copyright-policy.html.

Notice
Statements and opinions expressed in the chapters are these of the individual contributors and not necessarily those of the editors or publisher. No responsibility is accepted for the accuracy of information contained in the published chapters. The publisher assumes no responsibility for any damage or injury to persons or property arising out of the use of any materials, instructions, methods or ideas contained in the book.

First published in London, United Kingdom, 2020 by IntechOpen
IntechOpen is the global imprint of INTECHOPEN LIMITED, registered in England and Wales, registration number: 11086078, 7th floor, 10 Lower Thames Street, London, EC3R 6AF, United Kingdom
Printed in Croatia

British Library Cataloguing-in-Publication Data
A catalogue record for this book is available from the British Library

Additional hard and PDF copies can be obtained from orders@intechopen.com

Mobile Computing
Edited by Jesus Hamilton Ortiz
p. cm.
Print ISBN 978-1-78984-939-4
Online ISBN 978-1-78984-940-0
eBook (PDF) ISBN 978-1-83880-550-0

For EU product safety concerns:
IN TECH d.o.o., Prolaz Marije Krucifikse Kozulić 3, 51000 Rijeka, Croatia, info@intechopen.com or visit our website at intechopen.com.

We are IntechOpen, the world's leading publisher of Open Access books

Built by scientists, for scientists

4,800+
Open access books available

122,000+
International authors and editors

135M+
Downloads

151
Countries delivered to

Our authors are among the
Top 1%
most cited scientists

12.2%
Contributors from top 500 universities

BOOK CITATION INDEX
CLARIVATE ANALYTICS
INDEXED

WEB OF SCIENCE™

Selection of our books indexed in the Book Citation Index in Web of Science™ Core Collection (BKCI)

Interested in publishing with us?
Contact book.department@intechopen.com

Numbers displayed above are based on latest data collected.
For more information visit www.intechopen.com

Meet the editor

Jesus Hamilton Ortiz, DEA, PhD is a professional with wide experience as a professor and researcher in Computer Engineering Telecommunications and Mathematics. He is an international reviewer of recognized journals and an expert editor in ad hoc networks, mobile networks, computer technology, telecommunication networks, wearables, Industry 4.0, drones swarms, and algorithms. He has published more than 100 articles and seven books with more than 300,000 downloads. Dr. Ortiz is also a thesis director for undergraduates and postgraduates in telematics, computer, telecommunication, and electronic engineering programs. Currently, he is a professor at UNAD and CEO of CloseMobile R&D.

Contents

Preface **XI**

Section 1
Wireless **1**

Chapter 1 **3**
Wireless Communications Challenges to Flying Ad Hoc Networks (FANET)
by Miguel Itallo B. Azevedo, Carlos Coutinho, Eylon Martins Toda, Tassio Costa Carvalho and José Jailton

Chapter 2 **23**
An Overview of Query- Broadcasting Techniques in Ad Hoc Networks
by Naeem Ahmad and Shuchi Sethi

Chapter 3 **37**
Importance of Fifth Generation Wireless Systems
by K. Sakthidasan Sankaran, G. Ramprabu and V.R. Prakash

Section 2
Mobile Networks **49**

Chapter 4 **51**
Softwarization in Future Mobile Networks and Energy Efficient Networks
by Thembelihle Dlamini

Chapter 5 **65**
Localization Enhanced Mobile Networks
by Salman Al-Shehri, Pavel Loskot and Michael J. Hirsch

Section 3
Computing **85**

Chapter 6 **87**
Architecture and Operation Algorithms of Mobile Core Network with Virtualization
by Larysa Globa, Svitlana Sulima, Mariia Skulysh, Stanislav Dovgyi and Oleksandr Stryzhak

Chapter 7 109
Mobile Distributed User Interfaces
by Erika Hernández-Rubio, Amilcar Meneses-Viveros and Sonia G. Mendoza-Chapa

Chapter 8 123
Metabolic Health Analysis and Forecasting with Mobile Computing
by Zsolt P. Ori

Chapter 9 149
Energy Consumption Model for Green Computing
by Jesus Hamilton Ortiz, Fernando Velez Varela and Bazil Taha Ahmed

Preface

This book presents a vision of the present and future of mobile computing. It identifies and examines the most pressing research issues in the field. Comprising chapters written by leading researchers and academics, this book covers such topics as Flying Ad-Hoc Networks (FANETs), Vehicular Ad-Hoc Networks (VANETs), 5G, energy-efficient networks, localization in mobile networks, algorithms of mobile core networks, user interfaces, metabolic health analysis, and others.

The book is organized into the following three sections:

1. Wireless

2. Mobile Networks

3. Computing

Mobile computing is likely to play an important role in the future of society. This book presents the state of the art in mobile computing, wireless, and future applications. As such, this volume is suitable as a text for graduate students and professionals in the industrial sector and general engineering areas.

Jesus Hamilton Ortiz
CloseMobile R&D,
Spain

Section 1

Wireless

Chapter 1

Wireless Communications Challenges to Flying Ad Hoc Networks (FANET)

Miguel Itallo B. Azevedo, Carlos Coutinho, Eylon Martins Toda, Tassio Costa Carvalho and José Jailton

Abstract

The increasing demand for Internet access from more and more different devices in recent years has provided a challenge for companies and the academic community to research and develop new solutions that support the increasing flow in the network, applications that require very low latencies and more dynamic and scalable infrastructures, in this context the mobile ad hoc networks (MANETs) emerged as a possible solution and applying this technology in unmanned aerial vehicles (UAVs) was developed the flying ad hoc networks (FANETs) which are wireless networks independent, its main characteristics are to have high mobility, scalability for different applications and scenarios and robustness to deal with possible communication failures. However, they still have several constraints such as limited flight time of UAVs and routing protocols that are capable of supporting network dynamics. To analyze this scenario, two simulations were developed where it was possible to observe the behavior of FANET with different routing protocols both during data transmission and video transmission. The results show that the choice of the best routing protocol must take into account the mobility of the UAVs and the necessary communication priority in the network.

Keywords: FANET, UAV, routing, protocols, QoS, QoE

1. Introduction

The mobile ad hoc networks (MANETs) have evolved significantly in recent decades, where the differential of this type of network is the independence of a centralized infrastructure to organize flow, a router or switch, for example, so if any device present in this network is disconnected or damaged, it can automatically adjust to a new topology, and the routing tables are updated [1].

From the freedom provided by the MANETs, several other devices have started to be connected to the Internet in addition to cell phones, tablets, and notebooks, so that together with wireless sensor networks (WSN), they can play an essential role in the development of applications aimed at Internet of things (IoT) [2].

Among these devices, stand out cars, other types of cars and the highway itself in which they are, i.e., using a MANET network to connect them is the first step toward the creation of autonomous vehicles, vehicular ad hoc networks (VANET), for further study, see [3–5].

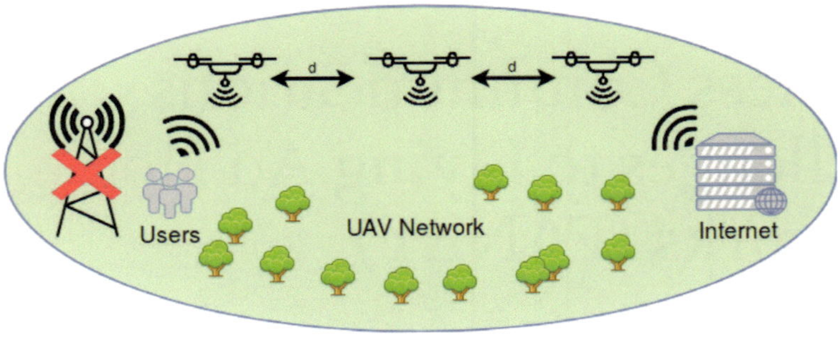

Figure 1.
FANET replacing LTE antenna.

They represent a major advance in mobile network technology in general, since several routing protocols have been developed or adapted for this type of network [6]. In addition, maintaining the quality of service (QoS) even at the high speed that the vehicles reach is one of the main challenges to be overcome in this area.

More recently, because of the popularization and cost reduction of unmanned aerial vehicles (UAVs), flying ad hoc networks (FANETs) [7–11] have emerged, which are networks composed only of aerial devices that can communicate between UAV-to-UAV and other ground-based UAV-to-ground devices. According to [12] UAVs can be divided into UAVs of high-altitude and long-range and medium-range, small drones and mini drones, the first two being military. In this chapter only the small drones and mini drones will be addressed.

The FANETs can operate either independently and by transmitting the flow received from land-based devices to a remote server or can also support other types of networks, for example, via satellite or cellular, if they are overloaded or unavailable as shown in **Figure 1**.

Therefore, such technology can play an essential role in the next generation of cellular networks, offering future support to 5G networks [13, 14], which it will reach very high speed and minimum latency, but due to the higher frequency in which they operate, the range is limited in relation to 4G. Thus, FANETs present themselves as a low-cost, scalable solution for maintenance and expansion of the Internet infrastructure worldwide, and it is essential to research, simulate, and validate their utility in different applications, taking into account the limitations of both UAVs and network itself.

In this chapter the main challenges for the implementation of FANETs in Section 2 will be addressed; in Section 3 the applications that can use the FANETs will be discussed and classified according to their characteristics, and Section 4 will address several available routing protocols and their advantages and limitations. Finally, Section 5 will demonstrate simulations of FANET behavior in different scenarios and routing protocols with data transmission and video.

2. Challenges

Despite several advances in recent years, FANET networks still have restrictions that may be critical to their operation depending on the application. The main one is energy consumption [15–17] because it limits the flight time of the drones, the speed of connection, and the range of the signal transmitted by them, so the challenges that need to be overcome to make FANET a reality involve primarily the search for solutions to these limitations, and in addition other factors can affect the performance of the network as the mobility and storage capacity of the drones. Possible solutions to these challenges will be discussed below.

Directional antennas: Most router antennas are omnidirectional, that is to say that the signal transmitted by them is sent equally in all directions, but when using these same antennas in drones, the results may not be very efficient as regards the quality of the antenna and energy consumption. Therefore, new antennas have been developed with beamforming technology [18, 19]; this change allows the transmitted signal to be directed to a specific area close to the UAV as shown in **Figure 2**. In this way the signal quality at the specific location is significantly better, and the energy consumption of the UAV is also reduced. However, it is still relatively a new technology and still needs to be better evaluated and implemented.

Mobility: One of the biggest differentials of UAVs is the high mobility and speed variation they have, which allows them to access hard-to-reach places and travel long distances in a short time, depending on the UAV model.

Thus, regardless of the mode of operation of the drones being both fully autonomous and controlled by a base station, it is necessary that critical information for the mobility of one or more drones is transmitted to the other drones in the network or to the base station as prevention alerts collision, GPS, flight time, environmental and climatic conditions, as well as the transmission of drone drive commands if they are controlled by a base station (**Figure 3**).

Routing protocols: Routing protocols are the brain of FANETs and control all flow both between UAVs and other devices connected to them, and although there are already several routing protocols available, these protocols sometimes cannot cope with mobility and the speed of the UAVs which causes a high rate

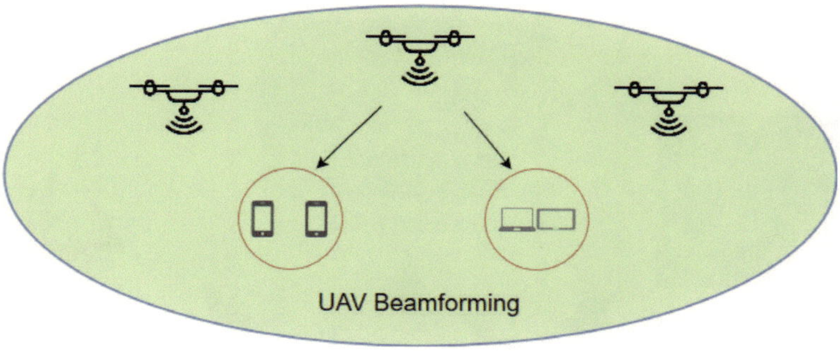

Figure 2.
UAV signal beamforming.

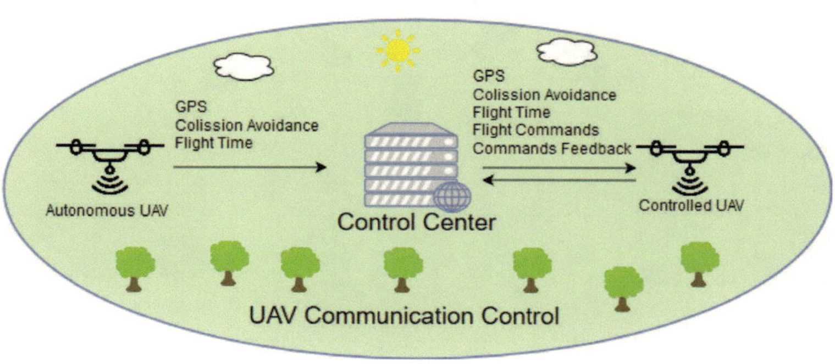

Figure 3.
UAV communication scheme.

of errors in the connection and until the drop of the network in certain cases. In this way, new protocols were developed with the focus on mobile ad hoc networks, with FANETs being one of them, and just like beamforming, these protocols are still being tested but at a later stage. Because the evaluation of routing protocols is one of the main research focuses in FANETs, it will be addressed in more depth in Section 4.

3. Applications

Due to its physical and architectural characteristics, there are several applications for FANETs. Some of them are mentioned in different scenarios.

3.1 Disaster monitoring

In some cases of disasters, a human being may encounter obstacles that prevent the analysis of the entire affected area. In this situation, it is possible to use FANETs to evaluate the scenario completely [7].

3.2 Monitoring of agricultural areas

There are several possibilities for the use of FANETs in agriculture such as complete crop evaluation, plant health analysis, and mapping of possible areas for planting expansion (**Figure 4**) [7].

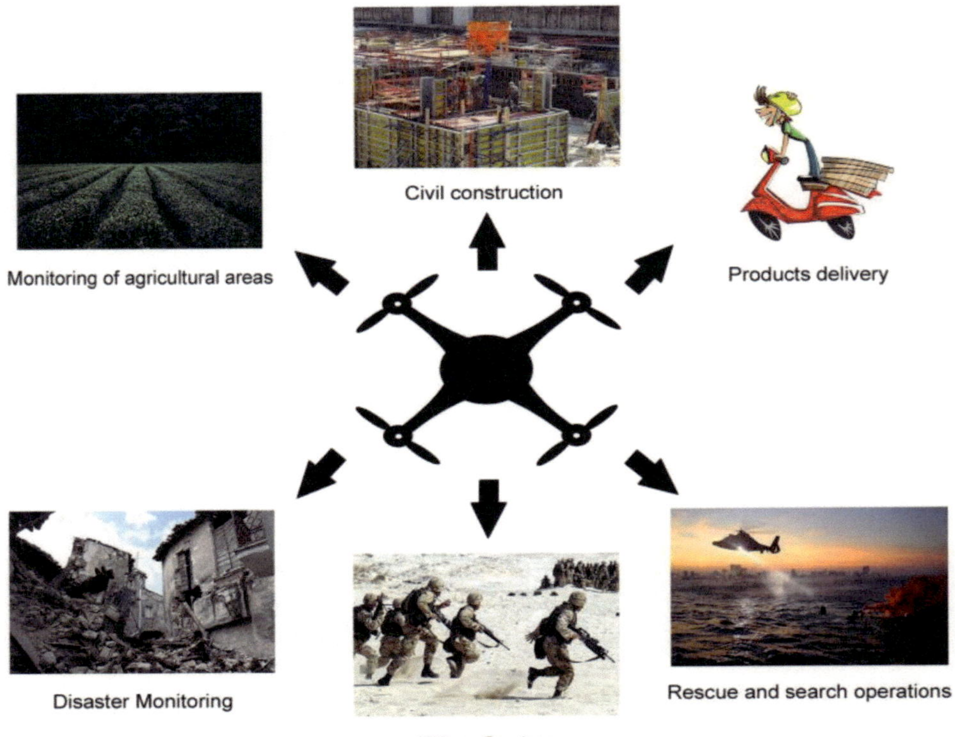

Figure 4.
FANET applications.

3.3 Search and rescue operations

In rescue situations where conventional mobile networks are damaged, the FANETs can be used to search for hostages in the affected area. And, because of the size of the UAVs, it is possible to access places in which a human would have difficulty [9].

3.4 Sensor networks

Sensor networks are mainly used for data collection and can be used together with FANETs in various situations [20]. Due to the ease of the UAVs to access any location without great difficulties, there will be an improvement in the performance of the networks when evaluating the scenarios in which they are applied [9].

3.5 Construction

With the use of FANETs, it is possible to analyze constructions, verify their progress and their quality, and also evaluate in advance the conditions of the environment to be used for the work in order to prevent possible calamities [7].

3.6 Product delivery

In order to reduce their costs and improve the quality of their services, some companies already idealize the use of UAVs for product delivery [7]. The service will be done autonomously through the implementation of an intelligent system in the UAVs [21].

3.7 Military service

The FANETs are widely used by military personnel primarily for communication between soldiers or between their barracks. It also can be used in civil operations to maintain the security of society [9].

4. Protocols

Due to some characteristics such as high mobility, the constant changes in network topology, unpredictable environmental and climatic factors, and energy consumption, the communication of nodes in FANETs becomes a challenge [7]. The nodes represent communication points in the networks, being computers, servers, or, in the case of FANETs, UAVs [22]. Therefore, in order for the network to have the desired performance, it is necessary that the routing protocols are adequate to handle various scenarios and conditions [23]. Here are the three types of protocols used in FANETs, which are proactive, reactive, and hybrid (**Figure 5**).

4.1 Proactive

Proactive protocols are those that update their routing tables at fixed time intervals. This feature allows the packets to be sent faster through the network because the nodes already know the changes in the routes [23]. The disadvantage of this type of protocol is the need for greater bandwidth to make constant updates [7]. The two main proactive protocols for FANETs are:

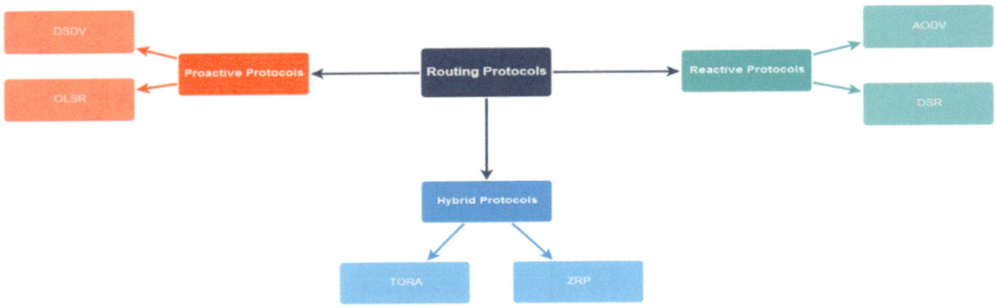

Figure 5.
Ad Hoc Routing Protocols.

4.1.1 Optimized link state routing (OLSR)

The OLSR is a specific protocol of ad hoc networks, and it has the main characteristic to select some nodes of the network to act as relays, called multipoint relay (MPR) [24]. The purpose of this mechanism is to avoid flooding in the network, caused by the excess of packets received by each node [25].

4.1.2 Destination-sequenced distance vector (DSDV)

In the DSDV protocol, in addition to the information of the network itself, the routing tables contain a sequence of numbers that changes according to the network topology [3]. These numbers are sent to all the nodes of the network, keeping them always updated, in order to avoid loops occurring between the nodes [23].

4.2 Reactive

Unlike the proactive ones, the reactive protocols only establish communication in the network when requested by the nodes [7]. Due to this feature, routing tables are only updated when there are packets to be sent [23]. Therefore, since it is necessary for the node to search the route before sending the package, the time to complete the entire process until the delivery is made becomes greater [9]. For use in FANETs, the following protocols are proposed:

4.2.1 Ad hoc on-demand distance vector (AODV)

The AODV protocol first makes the discovery of routes when necessary to send a packet and stores that route, just after the packet is sent to its destination [26]. In addition, because it is a mobile network protocol, if there is a connection interruption, the maintenance of routes starts to update the routing tables and thus maintains the communication between the nodes [24].

4.2.2 Dynamic source routing (DSR)

The DSR is a widely used protocol in multi-hop wireless networks, and it has the feature of its source node storing the entire route up to the destination node [24]. As the AODV, the DSR protocol also performs route discovery when there is a need for communication between two nodes to send packets [27]. And it performs maintenance of routes if there is a change in the topology of the network and communication is interrupted [24].

4.3 Hybrid

The hybrid protocols are the combination of reactive and proactive protocols, using the best resources of each, being used mainly for large networks [24]. They are based on the concept of zones, where proactive protocols are used for intra-zone routing (nodes within the same zone) and reactive routing protocols are used for inter-zone routing [23]. The main protocols of this type are:

4.3.1 Zone routing protocol (ZRP)

In this protocol, each node has a different zone, so the neighboring node zones overlap [24]. In intra-zone routing, proactive protocols are used to maintain the routes; due to this, if the source and destination nodes are in the same zone, packet sending is done immediately [23]. Already in the inter-zone routing, we use reactive protocols to find the routes and maintain them [23].

4.3.2 Temporarily ordered routing algorithm (TORA)

The TORA protocol is characterized by each node of the network maintaining only the routing information of its neighbors updated [23]. Because it is a highly adaptive protocol, its reactions to network topology changes are reduced in order to minimize the propagation of control messages, so the network does not look for new routes when there is no need [24]. The TORA mainly uses reactive protocols but also uses proactive protocols in some situations [23].

5. Performance evaluation

The performance of FANET networks can be analyzed through computer simulations. In order to correctly evaluate these network types, two scenarios were simulated to simulate different transmission types, and each one used three of the aforementioned routing protocols: AODV, OLSR, and DSDV [28].

The network simulator 3 (NS-3) software was used for the construction of simulation scenarios based on C++ language [29]. Each simulation has specific parameters that constitute different situations for this type of network. The details of each scenario were defined according to **Tables 1** and **2**.

The scenario of **Table 1** is to evaluate the performance of packet flow between UAVs and a server; this type of application is very common in a network of sensors for data collection or monitoring. In this scenario, all drones send data to the server at the same time and vice versa.

The scenario in **Table 2** seeks to evaluate the performance of a video transmission between a drone and a server; this type of application is widely used in disaster monitoring and rescue in locations inaccessible by the ground. In this scenario, a drone flies on a certain space while transmitting a video to the server on the ground.

5.1 Simulation parameters

Different types of simulation parameters were considered for each of the analyzed scenarios. In simulation 1, three network performance parameters were considered: packet received rate, delay, and network throughput. In simulation 2, only a few parameters were considered in relation to simulation 1, the received packet rate and the delay. However, the video transmission that occurs in this scenario requires specific quality of experience (QoE) parameters to be analyzed, such as structural similarity (SSIM), signal peaks (PSNR), and video quality metrics (VQM).

5.1.1 Packet delivery ratio

This parameter is defined by the ratio between the number of packets that are sent and the number of packets that are actually received at the destination. The final fee is displayed as a percentage based on total packets sent.

Simulation 1: UDP flow on a FANET network with UAVs	
Parameter	**Value**
Simulation time	60 s
Simulation area	400 × 400 × 400 m^2
UAVs' quantity	4
UAVs' speed	0–15 m/s
Mobility model	Gauss-Markov
Transmission range	40 m
MAC protocol	IEEE 802.11a
Routing protocols	OLSR, AODV, and DSDV
Transport protocol	UDP
Internet protocol	IPv4
Maximum transmission rate	100 kbps
Packet size	512 bytes

Table 1.
Simulation parameters for QoS.

Simulation 2: Streaming video on a FANET network with UAVs	
Parameter	**Value**
Time simulation	80 s
Simulation area	400 × 400 × 100 (L × P × A) m^2
UAVs' quantity	1
UAVs' speed	0–15 m/s
Mobility model	Gauss-Markov
Transmission range	40 m
MAC protocol	IEEE 802.11b
Routing protocols	OLSR, AODV, and DSDV
Transport protocol	TCP
Internet protocol	IPv4
Application	Evalvid
Video	st_highway_cif.st

Table 2.
Simulation parameters for QoE.

5.1.2 Delay

This parameter is defined by the amount of time a packet takes to traverse from the source to the destination in a transmission. Several factors directly influence this parameter, such as the routing protocols and the transmission rate of the network.

5.1.3 Throughput

This parameter is defined by the bandwidth of a transmission between two nodes of a network over time. It is generally expressed in bits/sec.

5.1.4 Structural similarity (SSIM)

It is a method used to measure the similarity between two images, determining the quality of the image received in the transmission [30]. In simulation 2, because it is a video, each frame receives a specific value. The SSIM metric was developed to extract structural characteristics of images, configuring the proximity between the pixels as a key factor to quantify the structures and to approach the visual quality of the images.

5.1.5 Peak signal-to-noise ratio (PSNR)

It is defined by the ratio between the signal power and the noise introduced by the transmission thereof [31]. In the case of video transmission, this noise influences the fidelity of the video representation at the destination. The PSNR is an approximation of human perception of the quality of reconstruction and is generally measured in decibels (dB).

5.1.6 Video quality metric (VQM)

This parameter is defined by the perceptual defect measurements of various video deficiencies, such as blur, rough motion, overall noise, block distortion, and color distortion. These measures are combined into a single metric that provides an overall quality forecast [32].

5.2 Results

In order to obtain the data, it is essential to organize and manipulate this data in a correct way, in order to analyze the performance of each parameter individually, considering the relation between them. In both scenarios, the flow monitor capture library [33] was used, which enabled the capture of the discrete-time performance parameters in each stream in the simulation scenario.

It was necessary to treat the data of the video transmission made by Evalvid in simulation 2 to extract the metrics of the network as delay and loss of packages. In order to obtain the PSNR and SSIM metrics, it was necessary to use the MSU Video Quality Measurement Tool (MSU VQMT), which reconstructs the transmitted video and compares it with the original Evalvid library video [34].

5.2.1 Simulation 1: UDP flow on a FANET network with UAVs

Comprising four UAVs and one server, the scenario simulates UDP packet flow between the server and the UAVs, which is initially in the server sense for UAVs (download) and UAVs for the server (upload).

Two graphs were generated for each analyzed parameter, one for upload flow and the other for download flow. Each displayed data represents the average of all the flows captured at that instant of time. After 60 seconds of simulation, all the performance parameters were analyzed.

5.2.1.1 Packet delivery

The received packet rate had similar results in both directions of flow. **Figure 6** shows the performance in the upload; the OLSR had the best performance, remaining above 70% in most of the simulation; on the other hand, the DSDV showed the worst performance, remaining below 40% in almost all the simulation.

In the download flow, the protocols had similar results throughout the transmission. The biggest difference between them can be seen in stability, especially the OLSR, which has remained stable close to 70% in most of the transmission. The AODV protocol exhibits poor performance in the initial simulation period, due to the reduced routing table of the protocol, which updates according to the need for transmission (**Figure 7**).

5.2.1.2 Delay

The resulting delay in the upload flow was noticeable only in the OLSR protocol, due to the common overhead of the proactive protocols [25]. On the other hand, in the download flow, the result was equivalent in all the protocols, concentrating the

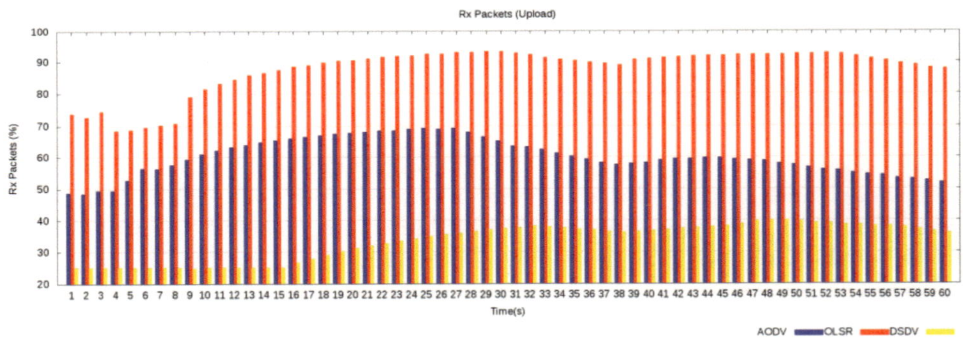

Figure 6.
Received packets in upload.

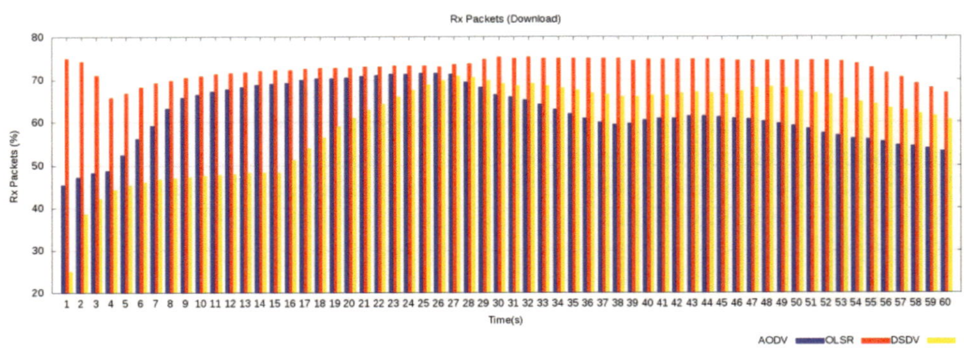

Figure 7.
Received packets in download.

greater delays at the beginning and the end of the transmission. Despite the high mobility of the drones, the delay had good results in download flow, in which no routing protocol exceeded 60 ms, as can be seen in **Figures 8** and **9**.

5.2.1.3 Throughput

The protocols performed differently depending on the flow direction. In the upload flow, shown in **Figure 10**, the OLSR had the highest average bandwidth between the protocols tested; on the other hand, the AODV and DSDV stabilized below half of the available maximum bandwidth, damaging the packet delivery, as previously seen in **Figure 6**. In the download flow, all the protocols have stabilized over time, since there being only one transmitter and several receivers, the packet flow became simpler, improving the overall performance of the protocols (**Figure 11**).

5.2.2 Simulation 2: streaming video on a FANET network with UAVs

Comprised of one UAV and one server, simulation 2 performs a video transmission between the server and the sleeping device. With the help of the Evalvid utility for NS-3, it was possible to analyze performance parameters and quality of experience (QoE) in all frames of the video, resulting in a detailed capture of protocol performance in this scenario. One was generated for each performance parameter, and four graphs for each QoE parameter.

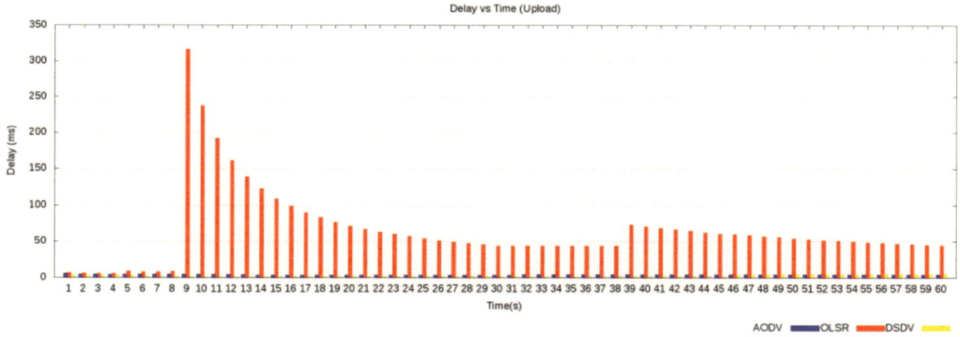

Figure 8.
Delay packets in upload.

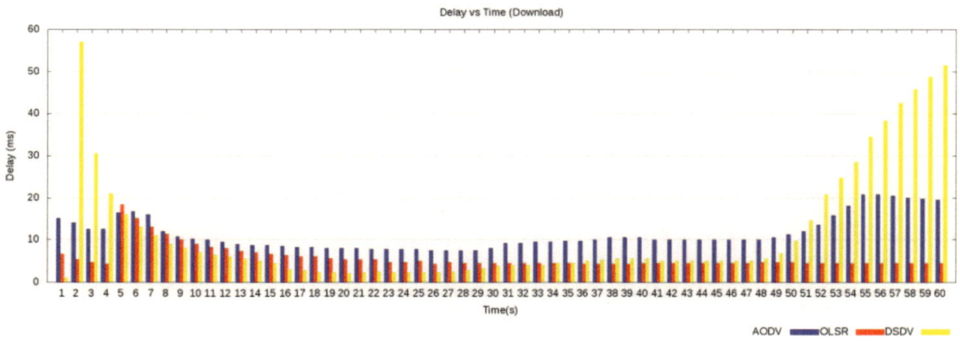

Figure 9.
Delay packets in download.

5.2.2.1 Packet delivery rate

The AODV and OLSR protocols had similar results along the transmission, with a higher performance in the first 17 seconds, followed by a significant decay due to the mobility of the FANET. The DSDV had the worst performance, as a result of an even greater decline, with no significant stability (**Figure 12**).

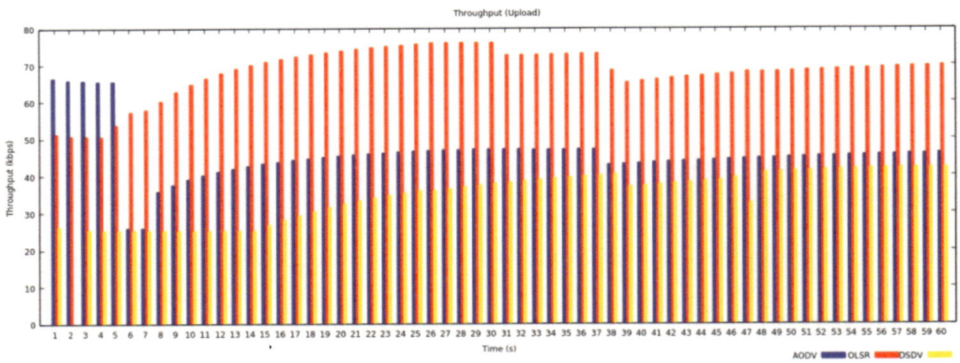

Figure 10.
Throughput in upload.

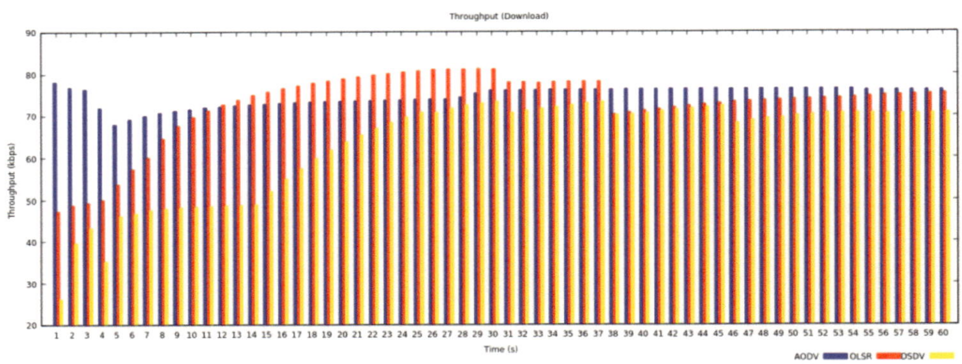

Figure 11.
Throughput in download.

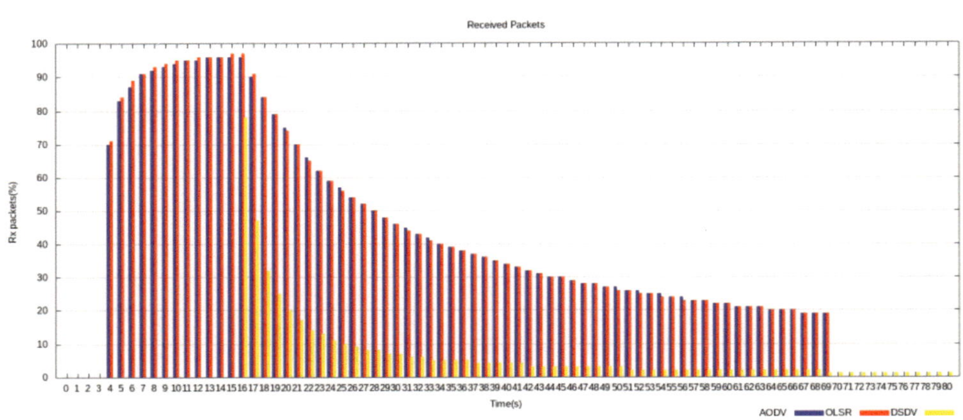

Figure 12.
Received packets in video application.

5.2.2.2 Delay

The delay was well-designed in all protocols, which stabilized throughout the transmission. The main factor that contributed to the result shown in **Figure 8** was the implementation of the Aar WiFi Manager function in simulation 2, an algorithm that works on the physical layer of the NS-3, which controls the transmission rate in the network [35]. With values below 3 ms in most of the transmission, the delay cannot be considered one of the reasons for packet loss (**Figure 13**).

5.2.2.3 Structural similarity (SSIM)

Figure 14 shows the mean SSIM in each protocol, which may have values of maximum 1 and have a minimum of 0.994 and a maximum of 0.998, highlighting a very low variation in transmission quality.

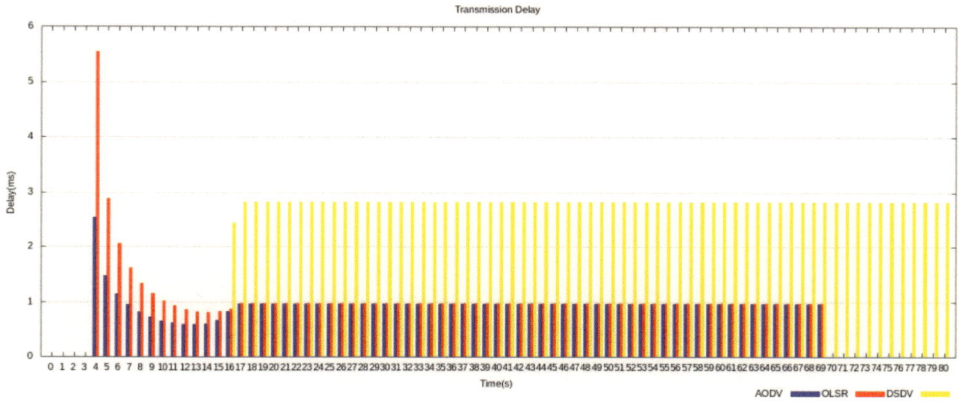

Figure 13.
Delay in video application.

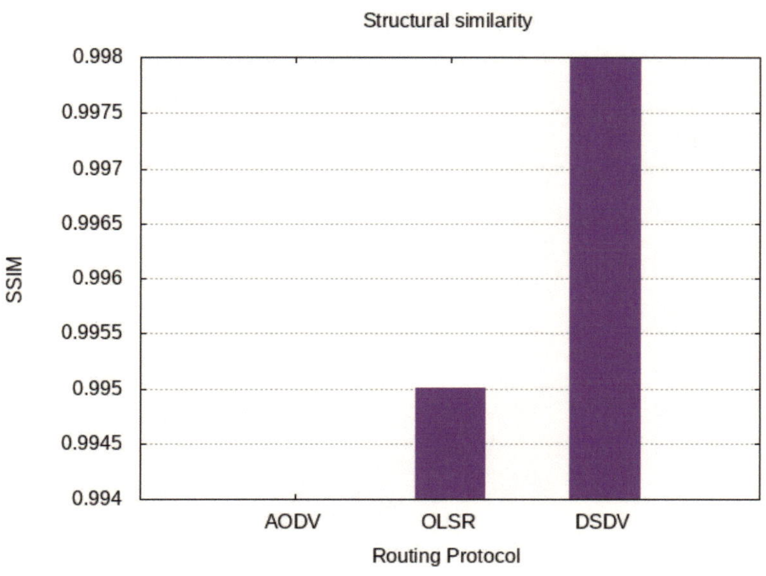

Figure 14.
SSIM average values.

Figures 15, 16, and **17** show the SSIM value in each frame of the video, further emphasizing the similarity in performance between the protocols. Thus, the routing protocols did not interfere directly in this metric; this was due to the direct

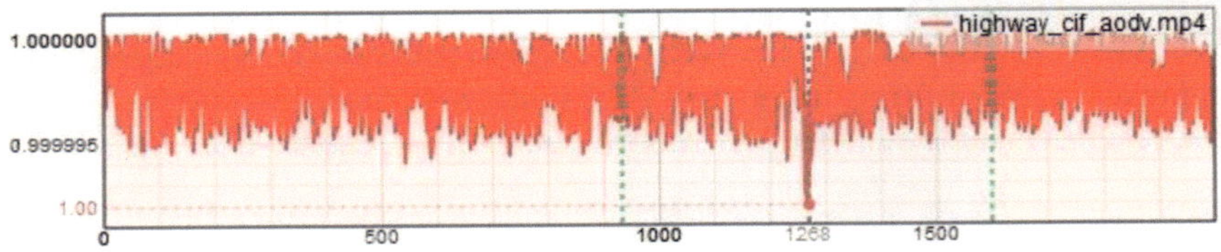

Figure 15.
SSIM to AODV protocol.

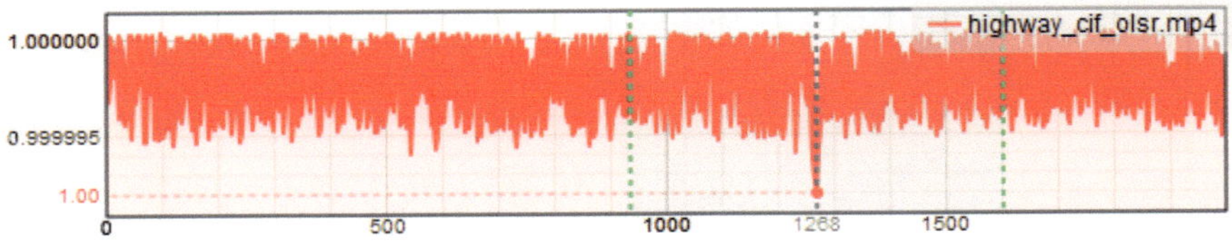

Figure 16.
SSIM to OLSR protocol.

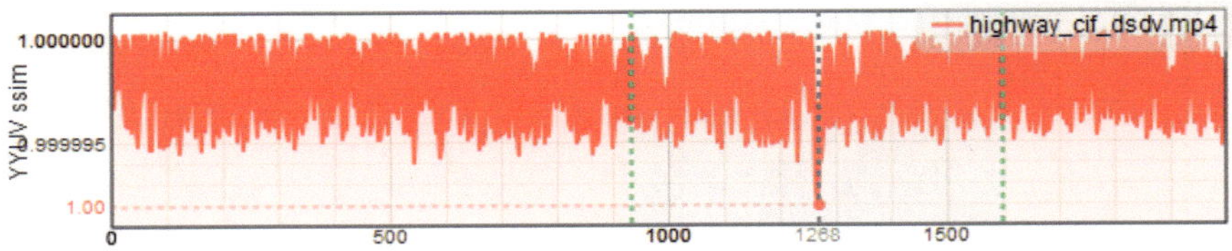

Figure 17.
SSIM to DSDV protocol.

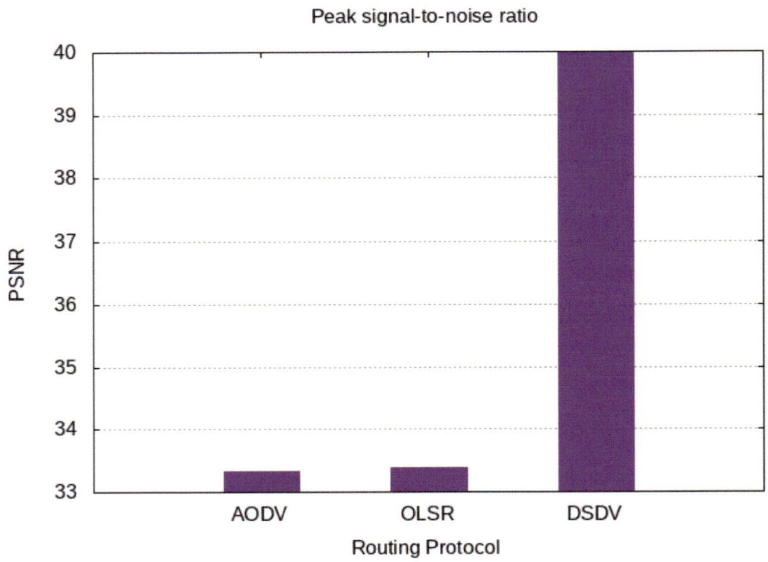

Figure 18.
PSRN average values.

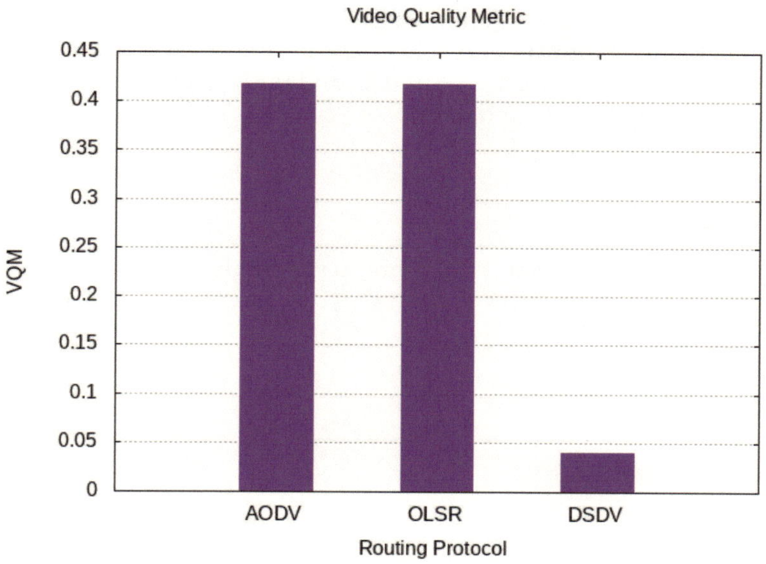

Figure 19.
VQM average values.

connection between drone and server during the simulation, without the need to perform much jumps in the network.

5.2.2.4 Peak signal-to-noise ratio (PSNR)

The PSNR shows the relationship between signal and video noise where a higher ratio means better quality. According to **Figure 18**, the DSDV obtained the best result, well above the other protocols, which are almost 7 dB difference, a significant value in this metric because it is a logarithmic scale.

5.2.2.5 Video quality metric (VQM)

The resulting VQM of **Figure 19** composes a correlation between seven different video quality parameters. The results show a low correlation coefficient in the AODV and OLSR protocol tests of slightly more than 0.4 between the original and transmitted videos. With the DSDV protocol, the resultant was even worse, with a value <0.05 correlation. In visual perception, this disparity results in a significant loss in video quality, something that was not shown in the other QoE parameters.

6. Conclusions

The evolution of the FANETs will allow a new range of application of this network, making several other devices connected to the Internet, such as sensors, cars, etc. Soon, FANETs will be essential for the construction of interim air networks. The applications discussed in this chapter have demonstrated the high flexibility of such networks, such as the use in rescue and monitoring, smart grids, etc. The challenges present in the FANETs are limited to problems in energy efficiency and routing protocols as seen in the simulations.

Due to the high mobility and flexibility of the UAVs, it is difficult to guarantee efficiency in all cases; simulations 1 and 2 have shown that proactive protocols are more efficient in scenarios that communicate with an onshore server but may not be the case in limited broadcasts or mobile server, which may have longer delays due to the frequent updating of the routing tables and the high mobility of the UAVs, which cause frequent loss of connection.

Acknowledgements

The authors also acknowledge the financial support by PROPESP/UFPA.

Author details

Miguel Itallo B. Azevedo, Carlos Coutinho, Eylon Martins Toda,
Tassio Costa Carvalho and José Jailton*
Federal University of Pará, Castanhal, Brazil

*Address all correspondence to: jjj@ufpa.br

IntechOpen

© 2019 The Author(s). Licensee IntechOpen. This chapter is distributed under the terms of the Creative Commons Attribution License (http://creativecommons.org/licenses/by/3.0), which permits unrestricted use, distribution, and reproduction in any medium, provided the original work is properly cited.

References

[1] Sharmila S, Shanthi T. A survey on wireless ad hoc network: Issues and implementation. In: 2016 International Conference on Emerging Trends in Engineering, Technology and Science (ICETETS), Pudukkottai; 2016. pp. 1-6

[2] Bruzgiene R, Narbutaite L, Adomkus T. MANET network in internet of things system. In: Ortiz JH, de la Cruz AP, editors. Ad Hoc Networks. London, UK: IntechOpen; 2017. DOI: 10.5772/66408

[3] Li F, Wang Y. Routing in vehicular ad hoc networks: A survey. IEEE Vehicular Technology Magazine. 2007;**2**(2):12-22. DOI: 10.1109/MVT.2007.912927

[4] Hartenstein H, Laberteaux LP. A tutorial survey on vehicular ad hoc networks. IEEE Communications Magazine. 2008;**46**(6):164-171. DOI: 10.1109/MCOM.2008.4539481

[5] Toor Y, Muhlethaler P, Laouiti A, La Fortelle AD. Vehicle ad hoc networks: Applications and related technical issues. IEEE Communication Surveys and Tutorials. 2008;**10**(3):74-88. DOI: 10.1109/COMST.2008.4625806

[6] Wahid I, Ikram AA, Ahmad M, Ali S, Ali A. State of the art routing protocols in VANETs: A review. Procedia Computer Science. 2018;**130**:689-694

[7] Cruz E. A comprehensive survey in towards to future FANETs. IEEE Latin America Transactions. 2018;**16**(3):876-884

[8] Bekmezci İ, Sahingoz OK, Temel Ş. Flying ad-hoc networks (FANETs): A survey. Ad Hoc Networks. 2013;**11**(3): 1254-1270

[9] Singh K, Verma AK. Flying adhoc networks concept and challenges. In: Khosrow-Pour M DBA, editor. Advanced Methodologies and Technologies in Network Architecture, Mobile Computing, and Data Analytics. Hershey, PA: IGI Global; 2019. pp. 903-911. DOI: 10.4018/978-1-5225-7598-6.ch065

[10] Zafar W, Khan BM. Flying ad-hoc networks: Technological and social implications. IEEE Technology and Society Magazine. 2016;**35**(2):67-74. DOI: 10.1109/MTS.2016.2554418

[11] Hayat S, Yanmaz E, Muzaffar R. Survey on unmanned aerial vehicle networks for civil applications: A communications viewpoint. IEEE Communication Surveys and Tutorials. 2016;**18**(4):2624-2661. DOI: 10.1109/COMST.2016.2560343

[12] Wang J, Jiang C, Han Z, Ren Y, Maunder RG, Hanzo L. Taking drones to the next level: Cooperative distributed unmanned-aerial-vehicular networks for small and mini drones. IEEE Vehicular Technology Magazine. 2017;**12**(3):73-82. DOI: 10.1109/MVT.2016.2645481

[13] Li B, Fei Z, Zhang Y. UAV Communications for 5G and beyond: Recent advances and future trends. IEEE Internet of Things Journal. April 2019;**6**(2):2241-2263. DOI: 10.1109/JIOT.2018.2887086

[14] Gapeyenko M, Petrov V, Moltchanov D, Andreev S, Himayat N, Koucheryavy Y. Flexible and reliable UAV-assisted backhaul operation in 5G mm wave cellular networks. IEEE Journal on Selected Areas in Communications. 2018;**36**(11):2486-2496. DOI: 10.1109/JSAC.2018.2874145

[15] Kerrache CA, Barka E, Lagraa N, Lakas A. Reputation-aware energy-efficient solution for FANET monitoring. In: 2017 10th IFIP Wireless and Mobile Networking Conference (WMNC); Valencia: 2017. pp. 1-6

[16] Bashir MN, Mohamad Yusof K. Green mesh network of UAVs: A

survey of energy efficient protocols across physical, data link and network layers. In: 2019 4th MEC International Conference on Big Data and Smart City (ICBDSC); Muscat. Oman: 2019. pp. 1-6

[17] Park J, Choi S, Hussen HR, Kim J. Analysis of dynamic cluster head selection for mission-oriented flying Ad hoc network. In: 2017 Ninth International Conference on Ubiquitous and Future Networks (ICUFN); Milan: 2017. pp. 21-23

[18] Biomo JMM, Kunz T, St-Hilaire M. Directional antennas in FANETs: A performance analysis of routing protocols. In: 2017 International Conference on Selected Topics in Mobile and Wireless Networking (MoWNeT); Avignon: 2017. pp. 1-8

[19] Xiao Z, Xia P, Xia X. Enabling UAV cellular with millimeter-wave communication: Potentials and approaches. IEEE Communications Magazine. 2016;**54**(5):66-73. DOI: 10.1109/MCOM.2016.7470937

[20] Akyildiz IF, Su W, Sankarasubramaniam Y, Cayirci E. A survey on sensor networks. IEEE Communications Magazine. 2002;**40**(8):102-114. DOI: 10.1109/MCOM.2002.1024422

[21] Job Selection in a Network of Autonomous UAVs for Delivery of Goods [Internet]. 1999. Available from: https://arxiv.org/ftp/arxiv/papers/1604/1604.04180.pdf [Accessed: 09 February 2019]

[22] Nó (redes de comunicação) [Internet]. 2017. Available from: https://pt.wikipedia.org/wiki/N%C3%B3_(redes_de_comunica%C3%A7%C3%A3o) [Accessed: 09 February 2019]

[23] Khan MA, Safi A, Qureshi IM, Khan IU. Flying ad-hoc networks (FANETs): A review of communication architectures, and routing protocols. In: 2017 First International Conference on Latest trends in Electrical Engineering and Computing Technologies (INTELLECT); Karachi: 2017. pp. 1-9

[24] Sahingoz O. Networking models in flying ad-hoc networks (FANETs): Concepts and challenges. Journal of Intelligent and Robotic Systems. 2014;**74**. DOI: 10.1007/s10846-013-9959-7

[25] Singh K, Verma AK. Applying OLSR routing in FANETs. In: 2014 IEEE International Conference on Advanced Communications, Control and Computing Technologies; Ramanathapuram: 2014. pp. 1212-1215

[26] Perkins CE, Belding-Royer EM, Samir Das. Ad-hoc on-demand distance vector (AODV) routing. In: Proceedings of IEEE WMCSA'99; New Orleans. LA: 1999

[27] Johnson DB, Hu Y-C, Maltz DA. The dynamic source routing protocol (DSR) for mobile ad hoc networks for IPv4. RFC. 2007;**4728**:1-107

[28] Farias MCQ. Video quality metrics. In: De Rango F, editor. Digital Video. London, UK: InTech; 2010. ISBN: 978-953-7619-70-1. Available from: http://www.intechopen.com/books/digital-video/video-quality-metrics

[29] Network simulator [Internet]. 2019. Available from: https://www.nsnam.org [Accessed: 02 February 2019]

[30] Structural similarity [Internet]. 2019. Available from: http://www.imatest.com/docs/ssim/ [Accessed: 06 March 2019]

[31] Peak signal-to-noise ratio [Internet]. 2019. Available from: https://www.semanticscholar.org/topic/Peak-signal-to-noise-ratio/11673 [Accessed: 06 March 2019]

[32] Vranješ M, Rimac-Drlje S, Grgić K. Review of objective video quality metrics and performance comparison using different databases. Imagine Communications. January 2013;**28**(1): 1-19. DOI: 10.1016/j.image.2012.10.003

[33] Flow monitor [Internet]. 2011. Available from: https://www.nsnam.org/docs/models/html/flow-monitor.html [Accessed: 02 February 2019]

[34] MSU quality measurement tool: Download page [Internet]. 2019. Available from: http://www.compression.ru/video/quality_measure/vqmt_download.html [Accessed: 02 February 2019]

[35] ns3: AarfWifiManager Class Reference [Internet]. 2019. Available from: https://www.nsnam.org/doxygen/classns3_1_1_aarf_wifi_manager.html#details [Accessed: 02 February 2019]

Chapter 2

An Overview of Query-Broadcasting Techniques in Ad Hoc Networks

Naeem Ahmad and Shuchi Sethi

Abstract

This chapter presents query-broadcasting techniques used to minimize expenses of the route discovery in ad hoc networks. A broad variety of such techniques have been proposed that improved the effectiveness and efficiency in various aspects of route discovery considering time and energy. Time-to-live based broadcast is the most common controlled flooding scheme widely used in routing protocols. One category of such techniques leveraged the routing history, while other category used broadcast repealing strategy to cancel the query-broadcast after successful route discovery.

Keywords: ad hoc networks, query-broadcasting, time-to-live, route discovery

1. Introduction

Freely moving mobile nodes arbitrarily create temporary structures called mobile ad hoc networks (MANETs). Low cost and ease of deploying attributes exist, owing to no requirement of preestablished infrastructure or centralized supervision for its configuration [1]. Each node in the network acts as a router with limited transmission range and is unable to directly communicate with nodes out of its transmission range. To communicate route discovery, using broadcasting query packet is employed, which can lead to flooding and broadcast storm [2, 3] and hence network congestion. This congestion takes toll on the energy consumption and average latency, thereby degrading the performance of the network. Research in the area has produced diverse set of techniques pertaining to packet broadcast expense control, keeping network congestion free.

This paper surveys and reviews all such proposed and adopted techniques under two major categories: confined broadcasting and unconfined broadcasting techniques as shown in **Figure 2**. In unconfined broadcasting techniques, source node broadcasting has no terminating condition, and each node [4] probes the set of selected neighbors based on metrics like weighted rough set (WRS) [5]. A cost-effective approach is that only participating neighbor nodes forward the query packets, while the rest discard the same [2].

These techniques have an edge of being reliable with assured success in finding optimal path in minimal time, thereby reducing packet duplication. The shortcoming of this category of techniques lies in its inability to control unnecessary retransmission of query packets despite known route. On the other hand, confined broadcasting set of techniques permits controlled flooding of the packets in a specified ring, thereby reducing congestion in networks. However, they compromise on the speed, and such approaches are very slow in finding the requested path [6]. Authors review all relevant and contemporary broadcasting approaches attempting to reduce such flooding expenses.

2. Route discovery in ad hoc networks

Route discovery is a process of finding optimal route (e.g., shortest, less congested, etc.) between two communicating nodes in the network. It is one of the characteristics of routing protocols, which may be reactive (on-demand) or proactive depending on the nature of routing protocols. The proactive route is made available in the table through periodic messages resulting in faster transmission. A few examples are OLSR [7], DSDV [1], etc. A class of power-aware routing protocols belongs to the proactive routing category. These protocols are loop free providing route in minimum time although the regular exchange of periodic messages congests the entire network using considerable storage space and draining energy of nodes [8]. Thus reactive routing protocols came into existence to reduce congestion and storage issues. Such protocols function on an on-demand basis (also called source-initiated routing protocols) initiating when the node needs to transmit, not requiring periodic transmission. Apparently, a large amount of battery power and bandwidth is saved. In reactive routing process, the source node broadcasts the query packet to the entire network, and intermediate nodes look in its cache for a route. If no route is available, re-broadcasting is done till the route to the destination is found. AODV [1] and DSR [9] use reactive routing.

To offset the limitations of each type, hybrid routing protocols emerged. These protocols use hierarchical approach to discover the route. It employs proactive approach within the proximity of the node and reactive approach between the proximity of nodes. Some examples are ZRP, IZRP, TZRP, AntHocNet, and HOPNET [10] and cluster-based routing protocols such as DWCA, DMAC, LEACH, and DTMNS [11].

In the above approaches, deploying flooding actually increases the cost to network by packet diffusion making routing expensive. To overcome its detrimental effects, various techniques are used at all levels from Mac layer to higher levels. Consequences of packet diffusion can be analyzed in AODV [1], Least Clusterhead Change (LCC), and ZRP [12] that are overcome in [10, 13] respectively. Similar broadcasting techniques were also proposed to reduce cost of packet diffusion [3]. The next section describes these approaches in detail.

3. Flooding of query packets

The process of disseminating the query packet to discover the optimal path (flooding) is the simplest form of broadcasting. Since every path is explored, the shortest and ideal path for effective transmission is guaranteed. Flooding was employed in many routing protocols such as AODV, OLSR, DSR, and DSDV.

Since packets traverse every outgoing edge of the directed graph shown in **Figure 1**, most nodes receive several copies of the same packet, and the intermediate nodes continue to forward the query packets to explore path, even after the route has been found, thus consuming a large bandwidth of the channel along with battery power of the participating nodes. This lowers the efficiency of the routing protocol. Two measures are taken to overcome the issue. First, the precautionary measure is opting for selective flooding, thus preventing redundancy of the packet

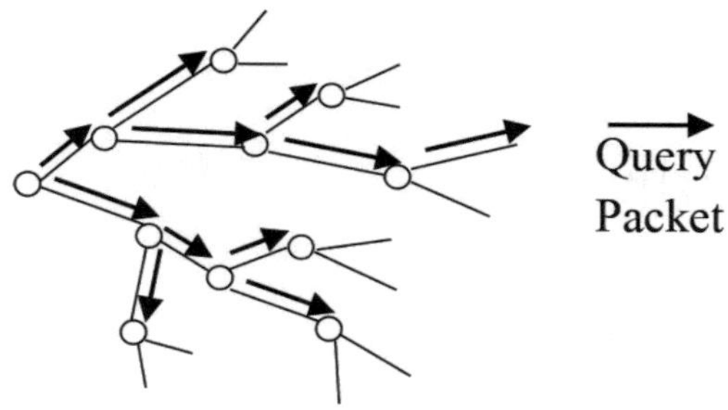

Figure 1.
Flooding of query packets in the network.

at intermediate nodes. Second, controlled flooding is employed to circumvent unnecessary propagation of query packets.

Assume that a graph represents a network. This network is a connected acyclic network where vertices of the graph are nodes and the edges between two nodes are connections. This network has N nodes creating an imaginary circle of diameter D. The average degree of each node is d (d > 2) representing the number of neighbor nodes.

Let PDC be the packet diffusion cost at a specified hop count, and it can be defined as

$$PDC = \frac{Total\ number\ of\ node\ at\ k\ hops}{Total\ number\ of\ node\ at\ k-1\ hops} \tag{1}$$

$$PDC = \frac{\sum_{i=1}^{k} d(d-1)^{i-1}}{\sum_{i=1}^{k-1} d(d-1)^{i-1}} \tag{2}$$

PDC of flooding excluding redundancy of packets at intermediate nodes for the entire network is given in the equation below:

$$PDC = \frac{d(d-1)^{R}-1}{d(d-1)^{R-1}-1} \tag{3}$$

where R is the radius of the network. By solving the above equation,

$$PDC = 1 + \frac{d-2}{1-1/(d-1)^{R-1}} \tag{4}$$

Assuming $a = d - 1$, we have the value of packet diffusion cost at R hop count given by the equation below:

$$PDC = \frac{a^{R}-1}{a^{R-1}-1} \tag{5}$$

Larger packet diffusion cost increases congestion in the network that leads to energy consumption problem, thus affecting network life calculated by the equation below:

$$EC_n = n \times E_r \tag{6}$$

where n is the number of nodes and E_r denotes the energy drained per node. In route discovery, energy is consumed in two ways: query-packet broadcast and reply-packet unicast. Let H_i be the number of nodes at ith ring and R be the radius of network. The energy consumption for flooding can then be shown as

$$EC_n = \sum_{i=0}^{H_R} E_i + E_{rrep} \tag{7}$$

where E_{rrep} is the consumed energy in unicasting reply packet. Following the aforementioned analysis, packet diffusion cost and energy drained for confined broadcasting techniques are calculated and shown in **Table 1**.

An optimization of blind flooding is broadcasting to intended nodes only. Broadcasting is essential to discover the choicest path along with other varied objectives. Some of them are listed below:

3.1 Reducing the flooding expenses

As already discussed, a main drawback of blind flooding is the broadcast storm [2] that congests the entire network. This congestion develops due to the redundant propagation of query packets. This undesirable circulation is reduced by the use of a suitable broadcast repealing technique.

3.2 Limiting the packet dropping

In ad hoc networks, multiple classes of congestion exist, leading to dropping of the packets. A traffic control technique is employed during the packet broadcast to estimate the traffic in the network. This enhances the reliability of the packet transmission [6, 14].

3.3 Optimizing the path length

End-to-end delay is the average time taken by the source node to transfer the packet successfully [15]. The length and traffic of the path determines the delay. Therefore, careful adoption of broadcasting technique optimizes the desired path.

3.4 Increasing reliability of the path

Reliability is determined by the stability of the path. Independent movement of the mobile nodes changes network topology which in turn causes link breakage.

Broadcasting techniques	Packet diffusion cost	Energy drained
LBA	$\frac{a^{2R}-1}{a^2-1}$	$\sum_{i=0}^{H_r} E_i + E_{rrep}$
TTL-ERS	$\frac{a^2(a^{2l}-1)}{(a^2-1)^2} - \frac{l}{a^2-1}$	$H_r * E_r + \sum_{i=1}^{H_r}\sum_{j=1}^{i} E_j + E_{rrep}$
BERS	$\frac{a^2(a^{2(k-l)}-1)}{(a^2-1)^2} - \frac{k-l}{a^2-1}$	$2\sum_{i=0}^{H_r} E_i + E_{rrep}$
BERS+	$\frac{a^{2k}-1}{a^2-1}$	$\sum_{i=0}^{H_r} E_i + E_{rrep}$
CMBERS+	$\frac{a^{2k}-1}{a^2-1} - C_R$	$\sum_{i=0}^{H_r} E_i + E_{rrep}$

Table 1.
Comparative study of different controlled flooding techniques.

Frequent link breakage decreases the reliability of the path [1, 9]. Therefore, broadcasting of the query packet is done in such a way that the packet can cover the least area which is sufficient to obtain the set of nodes with maximum battery life. Length of the path is another attribute determining the stable route with the shortest length.

3.5 Utilizing unicast and multicast modes

Although several routing protocols exist that work for unicast and multicast communication in MANETs, no routing protocol fits all scenarios due to varied nature of routing properties [1]. These properties are in turn dependent on broadcasting techniques. Consider a case, for example, there are five clients with each transmitting at 50 kbps in unicast mode. The group bandwidth turns out to be 250 kbps. While in multicast mode, the same load is experienced by 1 client to 250 clients. Thus the use of multicasting in confined broadcasting can reduce the cost of packet diffusion by customizing packet diffusion for group communication where the source node needs to find multiple routes at once for a set of nodes. In the unicast mode, unconfined broadcasting is useful along with the adoption of selective flooding.

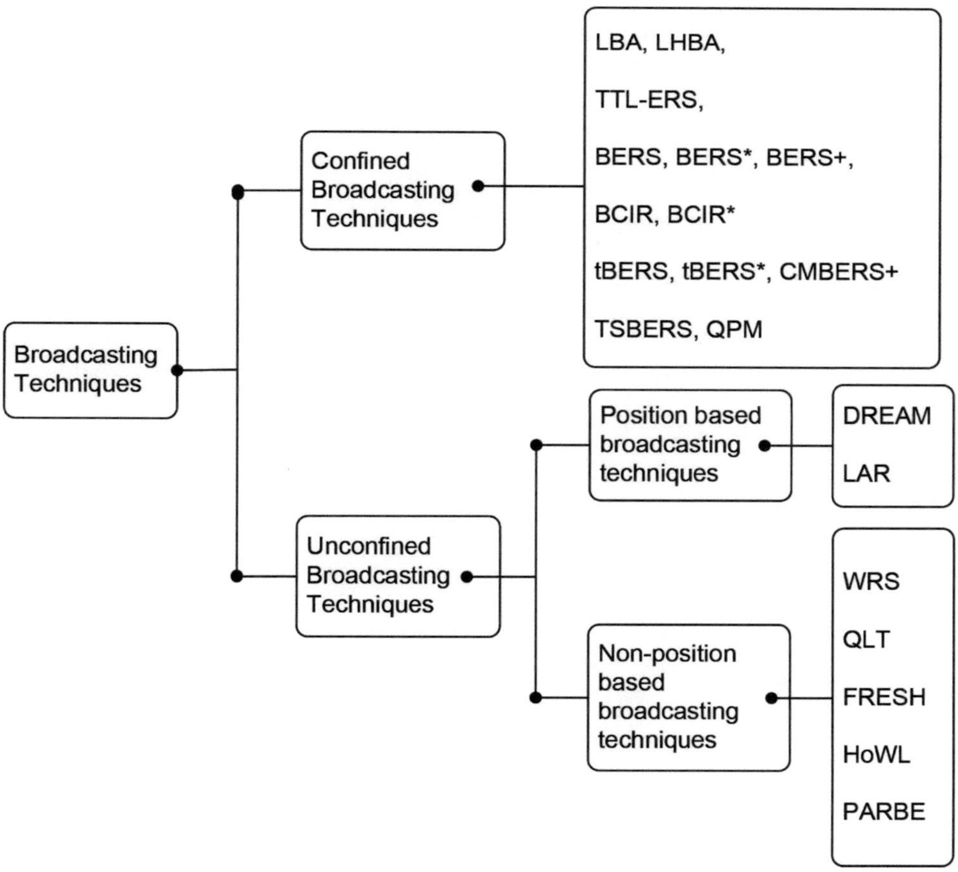

Figure 2.
Classification of broadcasting techniques.

4. Unconfined broadcasting techniques

Efficient and effective packet broadcasting during route discovery phase is pivotal to MANETs. As dynamic changes in topology occur, packet flooding gets costlier and poses the broadcast storm problem as well [2]. This situation worsens, when the source and destination nodes do not have record of previous communication. To prevent this situation, unconfined broadcasting techniques have been proposed. These approaches are based on the selective flooding, and thus blind flooding does not occur.

This is just like a modeled graph representing a network where initially all nodes are colored white. The source node determines a set of neighbor nodes based on attributes like position, knowledge, previous record, etc. The query packet is processed by nodes of the set, and such nodes are then colored either black or red as shown in **Figure 3**. This algorithm is iterative and the resultant set of participating nodes is obtained. As an example, WRS uses weight metric to choose forwarding set of nodes.

Similarly, position-based broadcasting techniques like LAR and DREAM, being scalable, reduce participating nodes by a considerable margin by exchanging location information in comparison to non-position-based techniques. But location-based techniques are not suitable where GPS signal reception is poor or inaccurate.

Another approach for reducing the congestion is knowledge-based technique. They have an advantage of not requiring any special device. These rely on previous communication, and with the increase in iterations, accuracy improves, and these techniques like HoWL and QLT [8] find a desirable route with much less effort than location-based techniques [2]. Comparative study of performance metrics is depicted in **Table 2**. These techniques prevent redundancy at intermediate nodes, thereby reducing congestion. But unconfined broadcasting does not have the capability to prevent unnecessary circulation of packets.

Anchor-based flooding employs primitive search in order to find the route. Anchor nodes are those nodes that have found the desirable route most recently. Every node maintains an encountering history consisting of the time of its last encounter with every other node. Source node searches the nearest anchor in its proximity using ERS [15]. Upon receiving route discovery packets, the anchor node informs the source node about itself and starts to search the next nearest anchor

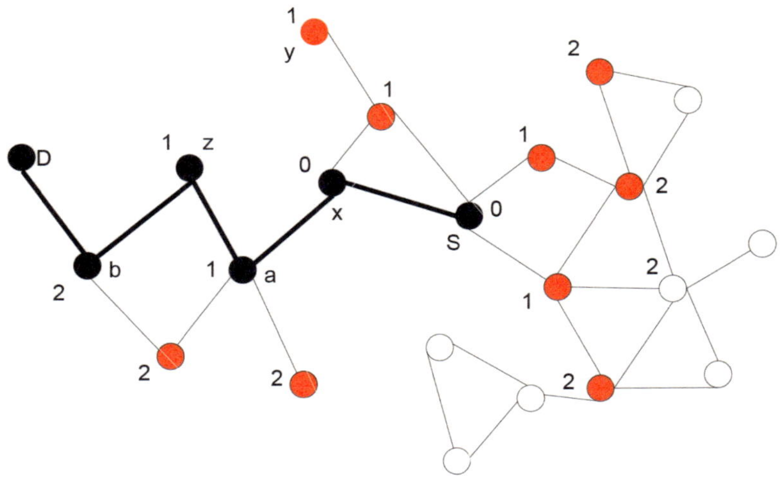

Figure 3.
A sample of the network representing the covered nodes in route discovery.

Broadcasting techniques	Path strategy	Type	Complexity	Hello message
FRESH	ABF	Proactive	O(n)	Yes
HoWL	RBF	Reactive	O(n)	No
WRS	NKBF	Reactive	$O(n^2)$	Yes
LAR	LBF	Reactive	O(n)	No
DREAM	LBF	Proactive	O(n)	Yes
QLT	RBF	Reactive	O(p + k)	No
PARBE	PBF	Reactive	O(n)	Yes

ABF, anchor-based flooding; NKBF, neighbor knowledge-based flooding; RBF, record-based flooding; LBF, location-based flooding; PBF, probability-based flooding; p, set of nodes lie in previous recorded route; k, threshold value.

Table 2.
Comparative study of the unconfined broadcasting schemes.

node. This practice is continued until the route node receives the query packet. These anchor nodes form the path from the source node to the destination node. An example of this approach is FRESH [17].

PARBE [18] is a probabilistic approach, aimed at reducing issues related to the route discovery process in AODV [1]. It helps in the reduction of unwanted searches during the route establishment process by considering the previous behavior of the network. Source node sends the query packet to only those intermediate nodes that have the probability to find the route to the destination. This probability is calculated using the previous record of requested path from the routing table. Unlike flooding, it does not require any freshet of the packet for route discovery.

5. Confined broadcasting techniques

The goal of confined techniques is preventing unnecessary circulation of query packets by limiting its hop count. Techniques like LBA [19], LHBA [20], revisiting-TTL ERS [21], blocking ERS [22], blocking ERS+ [15], BCIR [4], and tBERS [23] belong to this category. Chase-based strategy is used in almost all broadcasting techniques, revisiting-TTL ERS being an exception. LBA, for example, works in the following fashion: when a node starts route discovery, it broadcasts the query packet. On receiving, the destination node sends back a reply packet. After route discovery, the source node broadcasts the chase packets to terminate further propagation of the query packets. Limitations of high overhead were overcome in LHBA in a manner that single packet, based on reference bit, behaves as query, reply, or chase packet [19].

On the other hand, revisiting-TTL ERS shown in **Figure 4(a)** is an expanding ring search-based technique to control the flooding. It broadcasts the query packet periodically with increased time-to-live (TTL) value as attempts fail instead of using the chase packet to limit disseminated area of query packets. BERS, BERS*, and BERS+[24] are improvised versions of revisiting-TTL ERS depicted in **Figures 4(b)** and **(c)**.

The major advantage of chase-based approach is the guaranteed controlled flooding by canceling the packet broadcast at a specified hop in only one attempt though it causes channel overhead. Revisiting-TTL ERS, on the other hand, uses periodic packet broadcast to carry out the route discovery and increases the average latency and energy consumption and induces retransmission overhead. Other versions of this algorithm like BERS, BERS*, tBERS, tBERS*, BCIR, and BCIR* incur the

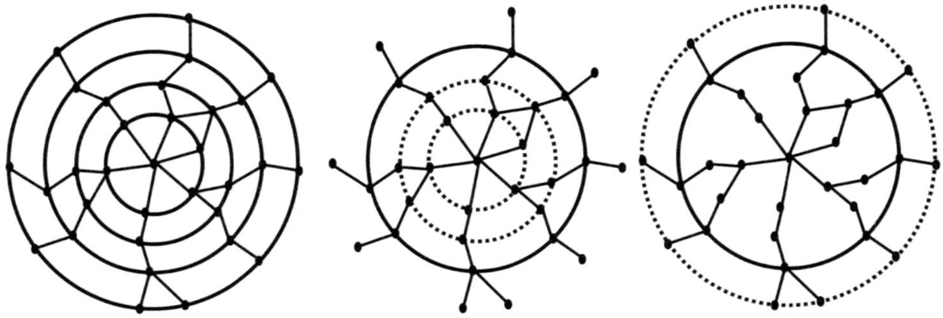

Figure 4.
Processing of ERS-based algorithm. (a) TTL sequenced-based ERS, (b) blocking ERS, and (c) blocking ERS+.

Broadcasting technique class	Unconfined broadcasting technique	Confined broadcasting technique
Method	Selective flooding	Controlled flooding
Packet disseminated area	Large enough area of the network to find the route; usually depends on the routing history and location as well, e.g., QLT, LAR, and DREAM	Small enough area of the network which depends on the predefined time-to-live (TTL) count
Control packets	No, prone to unnecessary propagation of query packets, e.g., WRS, HoWL	Yes, except using to control the further propagation of query packets, e.g., BERS, BERS+, tBERS
Applicable in	Proactive routing protocols where source node has link information of the whole network, which helps to prune the conveying intermediate nodes	Reactive routing protocols where the source node makes the first route discovery for any node
Storage requirement	Yes, increases as the number of nodes increases	No, however, some type of cache is used to track the predefined TTL value
Preferred for	Unicast mode	Multicast mode
Average latency	Very low, due to proactive nature	Higher due to added delay in the processing of query packet at each intermediate nodes
Periodic updates	Yes, require to gain previous routing information	Not required
Suitable	For small networks with high mobility	For large networks with slower to moderate mobility where no previous communication is available

Table 3.
Overall comparisons of broadcasting techniques.

same cost as conventional TTL-ERS in the worst-case scenario which is when predefined TTL value is small. This methodology is not adaptive in the case distance between the source and destination increases. BERS+ is adaptive to the mobility of the destination node and is best suitable where no previous communication exists. The drawback of BERS+ is that its broadcast termination is source initiated causing additional latency in the processing of control packets. However, BCIR, BCIR*, tBERS, and tBERS* apply destination-initiated broadcast termination approach offering higher retransmission efficiency over BERS, BERS*, and BERS+ [3, 6, 11, 14].

Apparently, there are a few more ways to tackle controlling the flooding. The method used to reduce the packet retransmission is cluster-based broadcast. The cluster heads and gateway nodes participate in packet retransmission, and other ordinary nodes remain silent. CBERS+ [16] is one such example. BERS+ is implemented in a destination-initiated manner, over a distributed clustered network that achieves scalability and broadcast termination. In highly dynamic networks, maintaining clusters is a difficult task as routing processing charges increase. Therefore, CMBERS+ is suitable for medium-sized networks with slow to moderate mobility with nodes that move in groups and where nodes are more likely to stay in groups. Overall comparisons of all techniques along with their features and applications are presented in **Table 3**.

6. Conclusion

In this chapter, almost all broadcasting techniques are reviewed under two categories: confined and unconfined. The unconfined broadcasting techniques, mainly derived from selective flooding, eliminate query packet redundancy at intermediate nodes in the route discovery, while confined broadcasting techniques reduce retransmission of query packets by controlling flooding. Most of the flat routing protocols employ only one broadcasting property of the two categories. Hybrid routing protocols, on the other hand, employ both properties by maintaining selective flooding within the proximity of node and controlled flooding between the proximity of nodes. Each technique has its merits and demerits. Unconfined approach of WRS and QLT and probabilistic approach offer simplicity in implementation where previous communications exist, while unnecessary flooding may be controlled with the use of a special device like NOVSTAR GPS to reduce conveying nodes as done in location-based algorithms DREAM and LAR. Though signal is weak, low accuracy due to atmosphere remains an issue. Confined broadcasting techniques save energy by employing strategies like TTL value to confine the region. Broadcasting techniques like TTL sequence-based ERS and its variants such as BERS and tBERS converge slowly for short predefined TTL value. BERS+ improves speed by introducing added delay after a maximum limit of TTL value. An enhanced version of CMBERS+ further increases the speed by issuing control packets at the route node. Moreover, scalability issue also has been resolved by dividing the network into distributed clusters. The advantage of these techniques over unconfined broadcasting techniques is that route discovery can be accomplished with controlled flooding when record of previous communication does not exist.

The central challenge in MANETs is exploring optimal path with minimal cost. A lot of research efforts have been devoted to the discovery of route using efficient and effective broadcasting technique. This chapter surveys shortcomings of the existing broadcasting techniques as well as discusses possible measures. Here are a few challenges that can be taken up in future research in the domain:

1. Blocking ERS+ introduced added delay after threshold to capture the query packets that slow down the route discovery after *k*th failed attempt. It can also be improved by reducing the added delay.

2. A comparative analysis of broadcasting techniques can be done in the clustered network which is still lacking in the majority of works.

3. Destination unreachability problem in LHBA can be removed to prevent the dropping of the gratuitous reply packet.

Moreover, Internet of Things (IoT) is a buzzword in the information and communications technology which covers a variety of routing protocols and their applications. In such scenarios, broadcasting techniques can play an important role in monitoring power theft, animals in the forest, automobiles with built-in sensors, etc. In this growing field, a controlled flooding will be required for multicast or group communication.

Notes/Thanks/Other declarations

Not Applicable.

Acronyms and abbreviations

ERS	expanding ring search
AODV	ad hoc on-demand distance vector
BERS	blocking expanding ring search
BERS+	enhanced BERS
BERS*	blocking expanding ring search*
tBERS	time-efficient BERS
tBERS*	time-efficient BERS*
BCIR	broadcast cancelation initiated on resource
MBERS+	modified BERS+
CMBERS+	cluster-based MBERS+
PARBE	probabilistic approach to reduce the broadcast expenses
LBA	limited broadcasting algorithm
LHBA	limited hop broadcasting algorithm
HoWL	hop-wise limited broadcasting
TSERS	two-sided ERS
QPM	query packet minimize technique
FRESH	fresher encounter search
QLT	query localization technique
DREAM	distance routing effect algorithm for mobility
WRS	weighted rough set
LAR	location aided routing

Author details

Naeem Ahmad[1*] and Shuchi Sethi[2]

1 Madanapalle Institute of Technology and Science, Madanapalle, India

2 Jamia Millia Islamia (A Central University), New Delhi, India

*Address all correspondence to: drnaeemahmad@mits.ac.in

IntechOpen

© 2019 The Author(s). Licensee IntechOpen. This chapter is distributed under the terms of the Creative Commons Attribution License (http://creativecommons.org/licenses/by/3.0), which permits unrestricted use, distribution, and reproduction in any medium, provided the original work is properly cited.

References

[1] Perkins CE, Royer EM. Ad-hoc on-demand distance vector routing. In: Mobile Computing Systems and Applications. 1999 Proceedings of the Second IEEE Workshop on WMCSA'99. IEEE. 1999. pp. 90-100

[2] Tonguz OK, Wisitpongphan N, Parikh JS, Bai F, Mudalige P, Sadekar VK. On the broadcast storm problem in ad hoc wireless networks. In: 3rd International Conference on Broadband Communications, Networks and Systems, BROADNETS 2006, IEEE. 2006. pp. 1-11

[3] Barjini H, Othman M, Ibrahim H, Udzir NI. Shortcoming, problems and analytical comparison for flooding-based search techniques in unstructured p2p networks. Peer-to-Peer Networking and Applications. 2012;**5**(1):1-13

[4] Lima R, Baquero C, Miranda H. Broadcast cancellation in search mechanisms. In: Proceedings of the 28th Annual ACM Symposium on Applied Computing, SAC '13. ACM; 2013. pp. 548-553

[5] Aitha N, Srinadas R. A strategy to reduce the control packet load of AODV using weighted rough set model for MANET. International Arab Journal of Information Technology. 2009. pp. 108-116

[6] Ahmad N, Hussain SZ. Analytical comparisons of query-broadcast repealing schemes in manets. Telecommunication Systems. 2018:1-13

[7] Clausen T, Jacquet P, Adjih C, Laouiti A, Minet P, Muhlethaler P, et al. Optimized link state routing protocol (OLSR). RFC 3626 (Experimental); 2003

[8] Maleki M, Dantu K, Pedram M. Power-aware source routing protocol for mobile ad hoc networks. In: Proceedings of the 2002 International Symposium on Low Power Electronics and Design, 2002. ISLPED'02. IEEE. 2002. pp. 72-75

[9] Johnson DB, Maltz DA, Broch J, et al. DSR: The dynamic source routing protocol for multi-hop wireless ad hoc networks. Ad-Hoc Networks. 2001;**5**: 139-172

[10] Haas ZJ, Pearlman MR. Protocol, the performance of query control schemes for the zone routing. IEEE/ ACM Transactions of Networking (TON). 2001;**9**(4):427-438

[11] Hussain S, Ahmad N. Cluster based controlling of route exploring packets in ad-hoc networks. In: Advanced Computing, Networking and Informatics, vol. 28 of Smart Innovation, Systems and Technologies. Springer International Publishing; 2014. pp. 103-112

[12] Haas ZJ, Pearlman MR, Samar P. The Zone Routing Protocol (ZRP) for Ad Hoc Networks, Internet Draft. Mobile Ad-Hoc Network (MANET) Working Group, IETF; 2002

[13] Kataria J, Dhekne PS, Sanyal S. Acrr: Ad hoc on-demand distance vector routing with controlled route requests. 2010. arXiv preprint arXiv:1005.0139

[14] Ahmad N, Hussain SZ. Broadcast expenses controlling techniques in mobile ad-hoc networks: A survey. Journal of King Saud University-Computer and Information Sciences. 2015;**28**:248-261

[15] Al-Rodhaan MA, Mackenzie L, Ould-Khaoua M. Improvement to Blocking Expanding Ring Search for Manets. Glasgow, UK: Department of Computing Science, University of Glasgow; 2008

[16] Castaneda R, Das SR, Marina MK. Query localization techniques for on-demand routing protocols in ad hoc networks. Wireless Networks. 2002;**8** (2–3):137-151

[17] Dubois-Ferriere H, Grossglauser M, Vetterli M. Age matters: Efficient route discovery in mobile ad hoc networks using encounter ages. In: Proceedings of the 4th ACM International Symposium on Mobile Ad Hoc Networking & Computing. ACM; 2003. pp. 257-266

[18] Preetha K, Unnikrishnan A, Jacob KP. A probabilistic approach to reduce the route establishment overhead in AODV algorithm for Manet. arXiv preprint arXiv:1204.1820; 2012

[19] Gargano L, Hammar M. Limiting flooding expenses in on-demand source-initiated protocols for mobile wireless networks. In: 18th International Parallel and Distributed Processing Symposium, 2004, Proceedings. IEEE. 2004. p. 220

[20] Zhang H, Jiang Z-P. On reducing broadcast expenses in ad-hoc route discovery. In: Distributed computing systems workshops. In: 2005. 25th IEEE International Conference on IEEE. 2005. pp. 946-952

[21] Chang N, Liu M. Revisiting the ttl-based controlled flooding search: Optimality and randomization. In: Proceedings of the 10th Annual International Conference on Mobile Computing and Networking. ACM; 2004. pp. 85-99

[22] Park I, Kim J, Pu I, et al. Blocking expanding ring search algorithm for efficient energy consumption in mobile ad hoc networks. In: WONS 2006: Third Annual Conference on Wireless On-demand Network Systems and Services. 2006. pp. 191-195

[23] Pu IM, Stamate D, Shen Y. Improving time-efficiency in blocking expanding ring search for mobile ad hoc networks. Journal of Discrete Algorithms. 2014;**24**:59-67

[24] Pu IM, Shen Y. Enhanced blocking expanding ring search in mobile ad hoc networks. In: 2009 3rd International Conference on New Technologies, Mobility and Security (NTMS), IEEE. 2009. pp. 1-5

Chapter 3

Importance of Fifth Generation Wireless Systems

K. Sakthidasan Sankaran, G. Ramprabu and V.R. Prakash

Abstract

Fifth generation wireless communications are denoted by 5G technology. 5G schemes are coming from first generation analog communication, 2G of Global System for Mobile communication (GSM), then 3G of Code Division Multiple Access (CDMA), after that fourth generation of long-term evaluation (LTE), and now fifth generation World Wide Wireless Web (WWWW). This research investigation presents issues, challenges, and the importance of 5G Wifi communication. In the 5G digital cellular network, the coverage area of the service providers is separated into small area called cells. All the audio, video, and image files are digitized and converted by an ADC (Analog to Digital Converter) and transmitted through stream of bits. 5G wireless devices are communicated using radio waves in a geographically reusable common pool of frequency band. Using wireless backhaul connection, the local antennas are connected with the internet/telephone network. Spectrum speed is substantially higher in millimeter wave. Hence, this was considered in this work.

Keywords: 5G, wifi communication systems, WWWW, challenges, issues and importance of 5G

1. Introduction

First, the goal of 5G network is to offer an extremely high speed data rate to enormous customers. Subsequently, to deploy a huge sensors in order to support abundant simultaneous connections, there must be a significant improvement inside the spectral performance of 5G [1] network when compared to 4G network. The telecommunication region has been introducing every 10 years a brand new generation of mobile networks in view that the creation of 1G. Introduction to any new cellular network requires new frequency assignment and a wide spectral BW (bandwidth). **Table 1** shows the progress of different telecommunication structures and their corresponding spectral bandwidth.

Other parameters like bit rate (higher peak), managing of concurrently connected devices, spectral performance, lesser battery intake, outage opportunity, higher bit rate, lower latencies, no. of supported gadgets, lower deployment cost and an additional dependable communication are predicted to be better in 5G. The anticipated deployment for this community is 2020.

Network will not suitable to support such increasingly more community usages is the major issue. In an effort to boom want to set up a flatter and greater dispensed community. The abundant file formats which includes all supported image, video, audio and information via NW (network) suggests that new source coding

IntechOpen

Network generation	Year of appearance	Spectrum value
4G	2012	<100 MHz
3G	2001	<20 MHz
2G	1991	<200 KHz
1G	1981	<30 KHz

Table 1.
Network generations and its spectrum value.

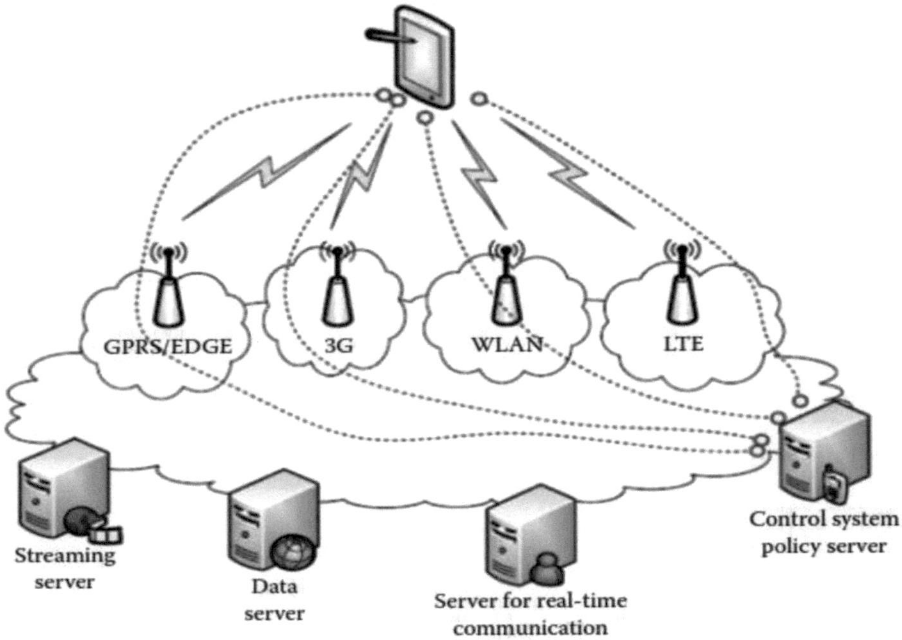

Figure 1.
5G structure for future networks.

along with H.264 is needed for sharing and shifting. Some other aspect that ought to be taken into consideration is using superior radio access networks (RANs) including heterogeneous networks, and complex methodologies for RAT which includes a new WWAN (wireless wide area network). In the future demand of 5G, improvement in technologies that are associated with transportation cell, network speed and interoperability should be increased. Typically, the optimization can be done on the programs, devices and network. 5G wifi method offers a huge excessive bandwidth by the way we use wifi devices. Further about 5G is, without any boundary limit, 5G will interconnect the complete world through an exclusive clever era. In order to offer an actual worldwide wireless web (WWWW), a new innovative idea of a multipath data course scheme is applied. To implement this kind of wifi globally network mixing is required. The final 5G architecture is designed in reality with multi-bandwidth path by collecting the current and network destiny. **Figure 1** depicts the structure, which incorporates the prevailing and destiny system.

Consequently, in such an actual 5G, code-division multiple access (CDMA), multicarrier code-division multiple access (MCCDMA), ultra wide band (UWB), orthogonal frequency-division multiplexing (OFDM), and Internet protocol version 6 (IPv6) will maintain the complete system. Because of such an in depth structure, by way of the use of 5G it will be potential to have super records talents and join limitless decision volumes and endless records broadcast. This potential needs that the implemented generation for access points and transfer in 5G has to offer an excessive connection for the community. Some other expectation of 5G is its potential to allocate net get admission to networks across the world at a clean velocity. The use of 5G, the furnished decision for a wifi community could be excessive and there may be big bandwidth shaping in both the direction. Capacity in remote diagnostics is an extremely good characteristic of 5G era. Through faraway management users will experience a network with speedy solutions.

2. Challenges and issues in fifth generation wireless systems

The maximum appealing objective for destiny attacker within the impending communications system of 5G can be the consumer instrumentation, accessing of a network, communication through mobile operator and outside IP networks. To assist the density protection problems and challenge in 5G system parts, we bequest cell structures which will have a sway on the future 5G communications systems have a tendency to gift adviser samples of viable threats and assaults explicit to those parts [2]. To receive these examples, we have a tendency to discover threats and assaults against by means of exploiting explicit capabilities with this latest communication principle. For the instance attacks, we have a tendency to in addition speak potential mitigation methods resultant from the literature that enables you for a roadmap nearer to an additional bigger counter measures.

2.1 Access networks

In fifth generation wireless systems, access points are anticipated to be notably complicated and heterogeneous, together with a couple of distinctive radio get right of entry to technologies and other advanced get admission to schemes, consisting of femtocells in order to guarantee the carrier availability. In non-existence of 4G for a while, the UE have to set up the connection over networks of 2G or 3G. Still, the reality that 5G network structures will help to inherit all the protection troubles of the fundamental entry to network [3].

During the transition from 4G to 5G communication, more desirable safety techniques must be carried out to counter prominent protection threats on 5G get right of entry to networks. To cope with this trouble, potential safety threats to the future 5G get right of entry to networks need to be first of all diagnosed. In this phase, awareness on attack presence on modern 4G access points and HeNB femtocells which can also be viable assaults at the 5G networks.

2.2 User equipment

In the fifth generation wireless systems era, User Equipment (UE), which includes powerful clever phones and pills, may be a completely important a part of our day by day existence. Such system will provide an extensive range of attractive functions to permit give up consumers to access an abundance of high-first-rate customized offerings. In any case, the normal creating acknowledgment of the fate UE, joined with the increased measurements transmission skills of 5G systems, the colossal selection of open working structures and the truth that the future UE will help a tremendous style of network choices are issues that provides the future UE an ideal objective for cyber-criminals. Aside from the customary SMS/MMS-based absolutely Denial of Service (DoS) assaults, the future UE can likewise be revealed to more noteworthy refined assaults started from portable malware with the goal that you can focus on each the UE and the 5G cell network. The unprotected running structures will permit stop clients to introduce programs on gadgets, no longer easiest from trusted yet additionally from suspicious resources. Therefore, cell malware, which will be ensured in bundles made to give off an impression of being blameless programming, will be downloaded and introduced on stop client's cell devices presenting them to numerous dangers. Portable malware can be intended to empower aggressors to exploit the put away private realities on the gadget or to discharge assaults contrary to different substances, which incorporates diverse UE, the cell get right of section to systems, the cell administrator's middle system and distinctive outer systems connected to the cell focus network. Henceforth, traded off future cell phones will now not exclusively be a peril to their clients, anyway likewise to the whole 5G cell network serving those [4].

2.3 External IP networks

In 5G wifi systems, the goal of DDoS attack is an external IP networks, wherein cellular back end generates traffic and finally transmits it to the goal over the core cellular network. Moreover, outside internet protocol systems, which incorporate organization systems, might be a delicate objective for being undermined by methods for malware through tainted cell contraptions getting to them. In this subsection, we blessing a delegate situation, in view of on [5], of how an association network might be undermined through the kindled 5G cell phone of a worker.

2.4 Core network for mobile operators

Because of their IP-primarily depends open structure, 5G versatile structures may be powerless to IP ambushes which are ordinary compared to the Internet. DoS attacks, that are a central shot at the Internet these days, will be blessing on the predetermination 5G correspondences structures concentrated on substances on the portable administrator's center network [2]. In any case, the 5G versatile administrator's middle network might be moreover influenced by DDoS assaults focused on outside elements, anyway moving their malevolent site guests over it. Potential ambushes include:

Mobile Operator's Core Network is targeted by DDoS attacks:

- Signaling Amplification
- HSS Saturation
- External units over a Mobile Operator's Core Network are targeted by DDoS attacks.

3. Importance of fifth generation

Few importances of fifth generation wireless systems are demonstrated in **Figure 2**. In the subsequent, all of the importances is intricate and highlights their function and importance for accomplishing fifth generation.

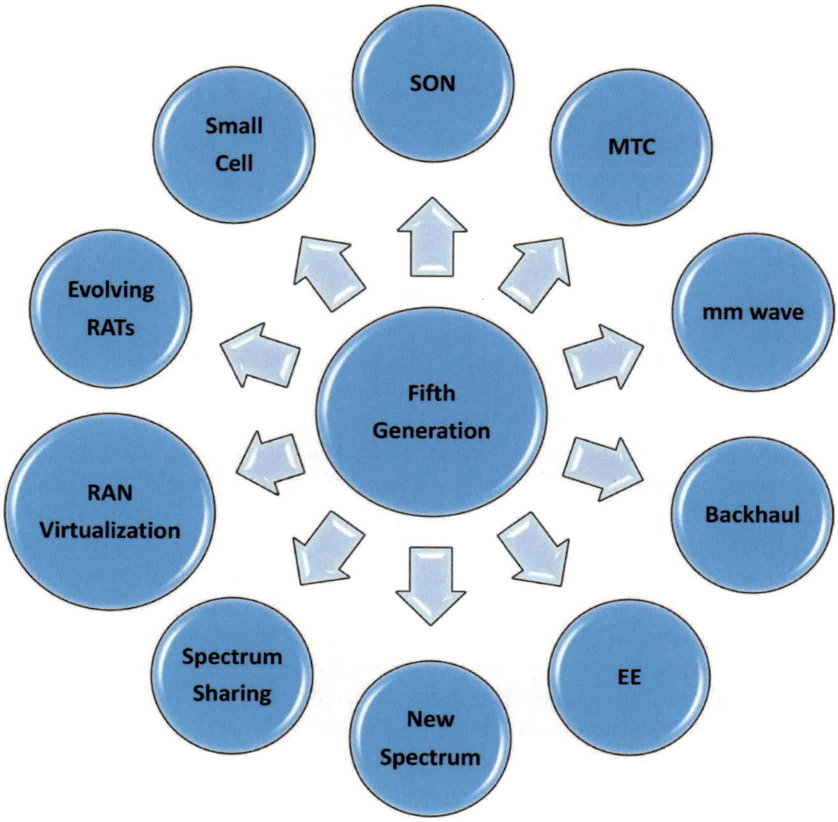

Figure 2.
Importance of fifth generation networks.

3.1 Existing RATs evolution

5th generation will rarely be a particular RAT, as a substitute it's miles possibly that it is going to be a combination of RATs along with the development of current methodology with new innovative plan complimented. In that capacity, the first and the most extreme intensely evaluated answer for adapt to the 1000× capacity crunch is the development of current RATs as far as SE, EE and inactivity, just as helping adaptable RAN sharing among more than one supplier. In particular, LTE needs to comply with help enormous/3-D MIMO to what's more exploit the spatial degree of freedom (DOF) by means of cutting edge multi-consumer bar shaping, to comparably decorate impedance crossing out and obstruction coordination capacities in a hyper thick small-cell sending situation. WiFi also wishes to conform to higher make the most to be had unauthorized spectrum. IEEE 802.11 ac, the modern-day enhancement of the WiFi era, can offer broadband wifi pipes with multi-Gbps facts charges. It uses the bandwidth extensively of hundred and sixty MHz on the much less polluted 5 GHz ISM band, using as much as 256 Quadrature Amplitude Modulation (QAM). It bolster concurrent transmissions up to 4 streams utilizing multi-person MIMO system [2]. The included shaft framing approach has helped the protection by methods for a few sets of significance, contrasted with its antecedent (IEEE 802.11n). At last, significant telecom organizations comprehensive of Qualcomm have as of late been running on creating LTE in the unlicensed range just as coordinating 3G/4G/WiFi handsets into a solitary multi-mode base station (BS) unit. In such manner, its miles expected that the upcoming UE will be shrewd adequate to lift the pleasant interface for making connection with the RAN fundamentally dependent on the QoS necessities of the currently running implementation.

3.2 Hyper dense small-cell deployment

When an additional EE is brought up in the device, hyper dense deployment of small-cell is every other challenging task to achieve the capacity in multiples of 1000. It is also called HetNet, Which in turn noticeably beautify the spectral efficiency (b/s/Hz/m2) of the region. Recently, many extraordinary ways are there to understand HetNet: (i) covering a cell device with small cells of identical technology with micro-, P.C.-, or femtocells; (ii) small-scale cells of various technology are masked in dissimilarity to simply the cellular one. The method is known as multi-tier HetNet, at latter point it has been denoted as multi-RAT HetNet.

On Qualcomm, a main organization in addressing 1000× potential undertaking via hyper dense small-cellular usage, has established that including small cells has the ability to scale the community nearly in a progressively increasing style, as depicted through diagram 3 [6]. That is, when small cells are increased then the capability also increased every time. Also, signal utilization and inter-mobile interference are increased when the cell length is reduced. To triumph over this downside, complex inter-cellular interference management strategies are wanted on the system stage together with complementary interference cancelation methods at the UEs. Enhancement of small-mobile changed into focus of LTE R-12, where the New Carrier Type (NCT) changed into delivered to manage small-scale cells by its host macro-cell. This lets in extra green manage plane functioning via the macro-layer at the same time by offering a maximum capacity and spectrally green records plane through the small-scale cells [7]. At last, reduction of cellular size enhances the network energy efficiency by keeping the community nearer to the UEs (**Figure 3**).

3.3 Self-organizing network

Self-Organizing Network (SON) usefulness is very important thing of 5G. SON benefits additional energy when the mass of little cells build. Practically 80% of the wifi site guests is created inside. To convey this enormous site guests, we need hyper thick small-cellular arrangements in houses—set up and kept up specifically with the guide of the clients—refractory administrators. Indoor little cells should be self-customizable and mounted in an attachment and recreational way [2]. Besides, the

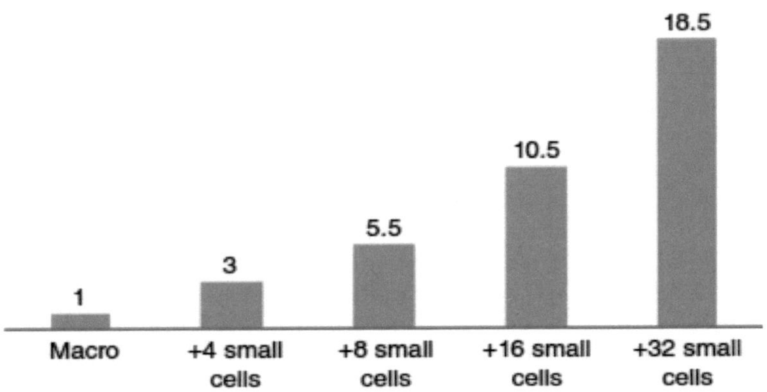

Figure 3.
Capacity scales linearly with the number of added small cells.

ability of SON is required to shrewdly adjust to the neighboring little cells to restrain inter-mobile obstruction. For exemplar, a little versatile can do that through self-governing synchronizing with the network and astutely altering its radio inclusion.

3.4 Machine type communication

Aside from people, interfacing cell machines is some other key issue of 5G. Machine type communication (MTC) is bringing programming up in which both one and both of the stop clients of the verbal trade meeting include machines. MTC forces two chief requesting circumstances on the system. To start with, the amount of gadgets that should be connected is generally enormous. Ericsson anticipates that 50 billion devices should be joined together inside the fate arranged society; the association imagines 'anything which can pick up from being connected may be connected' [8]. In the cell structure mechanism of a network, the diverse test forced with the guide of MTC is very quickening call for the management of actual-time and remote. It needs a truly low dormancy of considerably lower than an ms, so- alluded to as "tactile Internet" [9], managing 20× inactivity development from 4G to 5G.

3.5 Millimeter-wave rats development

The ordinary sub-three GHz range is to transform into progressively more clogged and the current RATs are drawing close to Shannon's capacity limitation. All things considered, examines on investigating cm- and mmWave groups for versatile correspondences has just been initiated. Despite the fact that the examinations on this era are still in its early stages, the output looks encouraging. It has three principle obstacles for mmWave portable correspondences. To begin with, the course misfortune is exceptionally better at those groups, contrasted with a standard sub-3GHz groups. Secondly, EM signals have a tendency to proliferate in the Line-Of-Sight (LOS) way, representing the connections of radio inclined and found to be obstructed by utilizing moving contraptions or individuals. Finally and yet importantly, the infiltration misfortune through the homes is considerably huge at these groups, shutting the exit way for indoor clients of RATs.

In spite of these confinements, there are heap gifts for mmWave interchanges. A significant measure of range is accessible in mmWave band; for instance, at 60 GHz, there might be 9 GHz of unlicensed range accessible. This is huge range, particularly the worldwide designated range for every cell innovation barely surpasses 780 MHz [10]. This measure of range can completely change portable interchanges through expert viding extremely-broadband remote pipes that can flawlessly stick the focused and the remote systems. Different advantages of mm Wave interchanges comprise of the little reception apparatus sizes (λ/2) and their little detachments, empowering many radio wire variables to be pressed in just one square centimeter. This in flip grants us to achieve exceptionally exorbitant bar shaping benefits in immensely little region, in such a way it can be connected at each the BS and the UE. Consolidating savvy staged cluster receiving wires, we will totally exploit the spatial level of opportunity of the wifi channel that may further improve the machine potential. At last, in light of the fact that the versatile station moves round, bar shaping loads can be balanced adaptively all together that the receiving wire shaft is always indicating the BS.

As of late, a manufacturing company boss is investigating mm Wave groups for cell correspondences, tried an innovation which could accomplish 2 Gbps realities cost with 1 km assortment in a city surroundings [11]. Besides, Professor Theodore Rappaport and his examinations bunch on the Polytechnic Institute of New York

University have approved that cell interchanges at 28 GHz in a thick metropolitan condition alongside Manhattan, NY, is reasonable with a portable size of 200 m the utilization of dual 25 dBi reception apparatuses, one in base station and second in the UE, it promptly sensible the utilization of cluster receiving wires and the shaft shaping strategy [10].

Finally, foliage misfortune for mmWaves is full-size and might confinement the engendering. Besides, mmWave transmissions may likewise encounter tremendous weakenings in the event of a substantial downpour for the reason that rain droplets are equivalent to the radio wavelength size (millimeters) and subsequently can reason dispersing. In this manner, a reinforcement cell contraption working in inheritance sub-3 GHz groups is most likely required as a piece of the mmWave answer [10].

3.6 Redesign of backhaul links

Updating the backhaul hyperlinks is consequent crucial trouble of 5th generation network. To improve the RAN in parallel backhaul interfaces should be reengineered to hold the huge client traffic produced in network cells. Something else, the backhaul connections will before long become bottlenecks, undermining the correct activity of the entire contraption. This issue increases more prominent energy on the grounds that the number of inhabitants in little cells increments [2]. Diverse verbal trade method can be thought by including optical fiber, mmWave and microwave. Explicitly, mmWave factor-to-point connections misusing exhibit radio wires with exceptionally sharp shafts are considered for solid self-backhauling in the absence of meddling with different cells or with the entrance hyperlinks.

3.7 Energy efficiency

EE is a challenging factor while growing 5G. Currently, Information and Communication Technology (ICT) expends as a decent arrangement as 5% of the power delivered far and wide and is responsible for about 2% of overall nursery fuel outflows—generally equivalent to the emanations made with the guide of the aeronautics venture. What issues more prominent is the truth that on the off chance that we do never again take any degree to decrease the carbon emanations, the commitment is anticipated to twofold by 2020 [12]. Consequently, it is important to seek after electricity-efficient format forms from RAN and backhaul connections for the UEs.

The advantage of power-green system layout is manifold. To begin with, it can assume a basic job in supportable improvement by lessening the carbon impression of a cell endeavor itself. Secondly, ICT in light of the way that the inside allowing development of the fate shrewd towns similarly can play an essential limit in cutting down the carbon impression of different parts. Third, it might expand the offers of portable administrators by bringing down their operational consumption by saving money with their quality bills. Before the last step, bringing down the 'Joule per bit' cost can save versatile contributions ease for the clients, permitting level rate evaluating paying little mind to the 10–100× measurements expense advancement foreseen through 2020. Last anyway no longer least, the battery life of the UEs can be extended, this has been observed by methods for the statistical surveying organization TNS [13] in light of the fact that the main measure of the overall population of the shoppers obtaining a wireless.

3.8 New spectrum allocation for 5G

Spectrum allocation of new gas wifi communications in the subsequent decade is an additional concern about 5G. The 1000× site visitors flow can infrequently be adjusted by means of most effective spectral performance enhancement or by using hyper-compaction. Actually, the main telecom groups which include Qualcomm and NSN consider that beyond technology modernization, extra bandwidth in multiples of 10 is wanted to fulfill the call for [2]. Allocating 100 MHz spectrums at 700 MHz and 400 MHz bandwidth at 3.6 GHz, in addition to the ability of numerous GHz bandwidth allocations in cm or mm wavelength to 5G can be focused the following WRC convention, classified by ITU-R in 2015.

3.9 Bandwidth sharing

Original spectrum management is highly time ingesting. Hence the utilization of available bandwidth is critical. Sharing bandwidth in various methods may be accepted to conquer the verified regulatory boundaries. Military radars are allocated with huge radio spectrums, where it is not completely applied over the period of 24 hours for all the days in a week or within the complete coverage area. In contrast, bandwidth purification is highly hard as a few spectrum are impossible to clean or will take very long time to clean it effectively; beyond that, the spectrum may be cleaned in a few locations however no longer in the whole state. Exactly, the Authorized/Licensed Shared Access (ASA/LSA) technique has been suggested via Qualcomm in order to obtain the advantage of spectrum in small-scale regions without interfering with the regime person [14]. This kind of allocation of spectrum can balance the spectrum cleaning slowly. This is well significant citing that as cell site visitor increase quickens, spectrum reforming becomes essential to smooth an already allotted bandwidth and ensuring its availability for 5G. Cognitive Radio ideas are also rechecked to mutually make use of certified and unlicensed spectrums. Finally, an innovative model to share the spectrum may be required as multi-tenant community operation and it will become sizeable.

3.10 Ran virtualization

The final but no longer least important is virtualization of RAN by enabling 5G, permits to share wifi infrastructure over multiple carriers. Network virtualization has to be driven from stressed out core community towards the RAN. For network virtualization, the intelligence wishes to be drawn out from the RAN and guided in a centralized method the usage of a software program mind, it can be accomplished in various community coats. Network virtualization can bring numerous blessings to the wifi area, inclusive of each Capex and Opex financial savings through multiple user network and sharing of the network system, stepped forward EE, increasing the required assets on-demand basis, by reducing the TTM (time to market) of modern offerings leads to expand the community agility, in addition to clean maintenance and rapid troubleshooting via improved transparency of the community [15–18]. Virtualization can serve converge for both wired and the wifi mesh by means of mutually coping with the entire community using primary orchestration set, also to improve the network performance. At last, multiple RANs assisting 3G, 4G or wireless may be accepted wherein exclusive interfaces using radio signals may be grew to become OFF or ON via CSPCU (central software program control unit) to enhance Quality of Experience (QoE) or Energy Efficiency (EE) of stop consumers.

4. Conclusion

The concept of fifth generation wireless communication technology WWWW is initiated from fourth generation LTE technique. Accordingly, fifth generation should create a significant divergence and include few extra services and traits to the global over fourth generation. Fifth generation must be gifted technology that communicates the globe with no edges. Consequently, in this article the main importance of fifth generation wireless communication systems are proposed and also the issues and challenges of fifth generation communication systems are also described. The major benefit of switching to 5G is convergence of multi-network functions in order to reduce the complexity, cost, power, rapid speed and incredibly low latency. Even though 5G provides better infrastructure for new business models, help to streamline communications, organizing to handle big data using its efficient transfer speed without going away from its core function. That is to serve as a mobile network. Also 5G enable us to explore technologies like virtual reality (VR) and augmented reality (AR). In the future, it will lend his hands on new innovations such as remote robotic surgery and personalized wearable health trackers. Not but the least IoT security will be a source of major investment in forthcoming years.

Author details

K. Sakthidasan Sankaran[1]*, G. Ramprabu[2] and V.R. Prakash[1]

1 Hindustan Institute of Science and Technology, Chennai, Tamil Nadu, India

2 New Prince Shri Bhavani College of Engineering and Technology, Chennai, Tamil Nadu, India

*Address all correspondence to: sakthidasan.sankaran@gmail.com

IntechOpen

© 2019 The Author(s). Licensee IntechOpen. This chapter is distributed under the terms of the Creative Commons Attribution License (http://creativecommons.org/licenses/by/3.0), which permits unrestricted use, distribution, and reproduction in any medium, provided the original work is properly cited.

References

[1] Arjmandi MK. "5G overview: Key technologies", chapter 2. In: Opportunities in 5G Networks. Boca Raton: CRC Press; 2016. pp. 19-32

[2] Rodriguez J. Fundamentals of 5G Mobile Networks. Wiley: John Wiley and Sons Ltd; 2015

[3] Piqueras Jover R. Security attacks against the availability of LTE mobility networks: Overview and research directions. In: Wireless Personal Multimedia Communications, IEEE, 16th International Symposium. 2013. pp. 1-9

[4] La Polla M, Martinelli F, Sgandurra D. A survey on security for mobile devices. IEEE Communications Surveys and Tutorials. 2012;**15**(1):446-471

[5] Li F, Peng W, Huang CT, Zou X. Smartphone strategic sampling in defending enterprise network security. In: Communications, IEEE International Conference. 2013. pp. 2155-2159

[6] Qualcomm 1000x: More Small Cells–Hyper-Dense Small Cell Deployments. 2004

[7] Hoymann C, Larsson D, Koorapaty H, Cheng JF. A lean carrier for LTE. IEEE Communications Magazine. 2013;**51**(2):74-80

[8] Ericsson. 5G Radio Access–Research and Vision, White paper. 2013

[9] Fettweis G, Alamouti S. 5G: Personal mobile internet beyond what cellular did to telephony. IEEE Communications Magazine. 2014;**52**(2):140-145

[10] Rappaport TS, Sun S, Mayzus R, et al. Millimeter wave mobile communications for 5G cellular: It will work! IEEE Access. 2013;**1**:335-349

[11] Pi Z, Khan F. An introduction to millimeter-wave mobile broadband systems. IEEE Communications Magazine. 2011;**49**(6):101-107

[12] Saghezchi FB, Radwan A, Rodriguez J, Dagiuklas T. Coalition formation game toward green mobile terminals in heterogeneous wireless networks. IEEE Wireless Communications. 2013;**20**(5):85-91

[13] TNS. 'Two-Day Battery Life' Tops Wish List for Future All-in-One Phone Device. 2005

[14] Qualcomm, 1000x: More Spectrum–Especially for Small Cell. 2013a

[15] Chowdhury NMK, Boutaba R. Network virtualization: State of the art and research challenges. IEEE Communications Magazine. 2009;**47**(7):20-26

[16] Goudos SK, Diamantoulakis PD, Karagiannidis GK. Multi-objective optimization in 5G wireless networks with massive MIMO. IEEE Communications Letters. 2018;**22**(11):2346-2349

[17] Salem AA, El-Rabaie S, Shokair M. Energy efficient ultra- dense networks based on multi-objective optimisation framework. IET Networks. 2018;**7**(6):398-405

[18] Hussain M, Rasheed H. March. Communication infrastructure for stationary and organized distributed smart meters. In: 2019 2nd International Conference on Communication, Computing and Digital Systems (C-CODE) (pp. 17-22). IEEE; 2019

Section 2

Mobile Networks

Chapter 4

Softwarization in Future Mobile Networks and Energy Efficient Networks

Thembelihle Dlamini

Abstract

The data growth generated by pervasive mobile devices and the Internet of Things at the network edge (i.e., closer to mobile users), couple with the demand for ultra-low latency, requires high computation resources which are not available at the end-user device. This demands a new network design paradigm in order to handle user demands. As a remedy, a new MN network design paradigm has emerged, called *Mobile Edge Computing* (MEC), to enable low-latency and location-aware data processing at the network edge. MEC is based on network function virtualization (NFV) technology, where mobile network functions (NFs) that formerly existed in the evolved packet core (EPC) are moved to the access network [i.e., they are deployed on local cloud platforms in proximity to the base stations (BSs)]. In order to reap the full benefits of the virtualized infrastructure, the NFV technology shall be combined with intelligent mechanisms for handling network resources. Despite the potential benefits presented by MEC, energy consumption is a challenge due to the foreseen dense deployment of BSs empowered with computation capabilities. In the effort to build greener 5G mobile network (MN), we advocate the integration of energy harvesting (EH) into future edge systems.

Keywords: softwarization, energy harvesting, soft-scaling, energy self-sustainability, forecasting

1. Introduction

The evolution towards a softwarized evolved packet core (EPC) is the driving force towards overcoming the challenges observed in current mobile networks (MNs) and set the way for handling the high data rate and ultra-low latency demands required by mobile users. This change avails the possibility of running network functions (NFs) in software, instead of proprietary hardware devices, and also it permits the possibility of dynamically scaling the network resources for a more robust network management in 5G and beyond as network complexity is reduced. Softwarization and virtualization of resources and services are undoubtedly among the main drivers of 5G and beyond 5G networks, as they will provide the mechanism for network management, that is, network flexibility and adaptability is guaranteed [1, 2], and also facilitate network maintenance and update all networks with ease.

To address the need for MN evolution, different design approaches have been investigated towards the state-of-the-art EPC architecture where vendors and

IntechOpen

researchers proposed the (i) grouping of the EPC functions [3–5], (ii) running the virtualized functions on clouds [6, 7], (iii) partitioning the network resources into network slices [8–11], which refers to an isolated set of (programmable) resources to enable NFs and services, and (iv) redesigning the network to be based on network function virtualization (NFV) technology [12, 13].

The architectural evolution involving the redesigning of the network based on NFV is currently appealing towards 5G and beyond, as it allows the virtualization of the mobile NFs and then placing them within the access network. This is motivated by the expected data explosion in the volume, variety, and velocity, generated by pervasive mobile devices and the Internet of Things [14] at the network edge (i.e., in close proximity to mobile devices, sensors, actuators, and connected things). This is coupled with the demand for stringent latency, requiring high computation resources which are not available at the end-user device. As a remedy, *Mobile Edge Computing* (MEC) has recently emerged to enable ultra-low latency, distributed intelligence and location-aware data processing closer to mobile users [2, 12, 13]. Undoubtedly, offloading to a powerful computation resource-enriched MEC server located closer to mobile users is an ideal solution.

Software-defined networking (SDN) and NFV are the emerging virtualization technologies that will enable flexibility and agility in MNs. SDN supports programmable interfaces to provide flexibility and agility on the network control management [14, 15], and NFV softwarizes the mobile NFs [16]. Both technologies limit the use of specialized hardware devices as they have been the limiting factor towards MNs evolution and the fast deployment of new services within the mobile space. The technologies can co-exist within the same network, where SDN employs a centralized approach on switching and routing elements [3], while NFV migrates the NFs out of dedicated hardware into software that is imported into general purpose hardware. In addition, these technologies are expected to facilitate the ease of efficient network management. The key to virtualization is that it enables *on-demand* or *utility* computing, a just in time resource provisioning model in which computing resources such as central processing unit (CPU), memory, and disk space are made available to applications only as needed and not allocated statistically based on the peak workload demand [17].

In light of the dense deployment pattern that is foreseen in 5G systems [18], the expected dense deployment of MEC servers and base stations (BSs) raises concerns related to energy consumption. Specifically, energy drained in BSs is due to the always-on approach (dimensioned for maximum expected capacity, yet traffic varies during the day), and in MEC servers, it is due to the computing-plus-communication processes associated with: (i) the running virtual entities [virtual machines (VMs) or containers] [19, 20]; (ii) the communication within the server's virtual local area network (VLAN) [21], and the presence of transmission (optical) drivers (fast tunable optical drivers for data transfer) within the MN infrastructure [22, 23]. Since energy consumption is a challenge within a BS system and the virtualized computing nodes (i.e., MEC servers), quantifying the energy consumption within computing platforms and transmission nodes is of great importance towards the development of robust energy management procedures. In addition, to address the carbon footprint emission and dependence in the powered grid, methods for using renewable (green) energy coupled with sustainable energy storage solutions are now receiving more attention than before. The motivation towards green energy is that the components in solar- and wind-based systems are usually modular, which makes the design, expansion, and installation of these types of systems for the BS sites very practical and feasible. This result in energy harvesting (EH)-powered BSs (EH BSs) and MEC (EH-MEC) systems.

1.1 Softwarization of mobile networks

Softwarization and virtualization is the driving force towards network evolution. Towards this end, researchers from industry and academia have presented different proposals towards the EPC architectural evolution. Their contributions result into fragmented inputs (see [2] for more details about EPC architectural proposals), with a unified goal of having an energy efficient EPC architecture for future MNs. The outcome is designs/proposals that are somehow overlapping in terms of the functions being softwarized, technologies used, and so forth.

1.1.1 Virtualization technologies and tools

The virtualization tools consist of the *hypervisor* and *docker engine*, and they are illustrated in **Figure 1**. Virtualization making use of the hypervisor is referred to as *hypervisor-based* virtualization, with the VMs as the computing resources or computer emulators, and for the environment using docker engine is referred to as *container-based* virtualization, with the containers as the computing resources. These different virtualization techniques share a similar architectural structure. However, the way in which each technique builds virtualized applications on top of the underlying supporting software is rather different.

A *hypervisor-based virtualization* simply isolates the operating system (OS) and applications from the underlying computer hardware [24]. This abstraction allows the underlying "host machine" hardware to independently operate one or more VMs as "guest machines" (also referred to as guest VMs), allowing them to share the systems physical computing resources, such as processing time, memory space, network bandwidth, etc. The hypervisor is the software that provides the environment in which the VMs operate. A new agnostic OS is generated to manage the underlying resources. Since the hypervisor sit between the actual physical hardware and the guest OS, it is also referred to as virtual machine monitor (VMM).

Container-based virtualization involves the use of containers as the computing resources within the virtualized computing platforms. Containers are abstraction units for isolating applications and their dependencies that can run in any environment. They can run on the same machine, on top of the docker engine, sharing the OS kernel with other containers. The docker engine is a software technology written

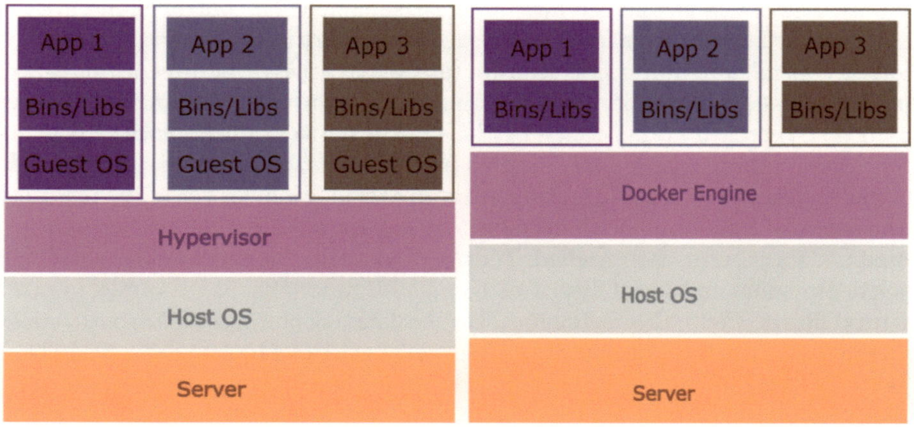

Figure 1.
An illustration showing the architectures of virtualization technologies: hypervisor-based virtualization (left) and container-based virtualization (right) [2].

in the Go programming language and it avails the environment for developing and running applications. It runs natively on Linux systems (in recent Linux kernels), where it uses Linux kernel features like namespaces, to provide a private, restricted view on certain system resources within a container (i.e., a form of sandboxing), and control groups (cgroups), a technology that limits an application to a specific set of resources (i.e., provides resource management, e.g., limits and priorities, for groups of processes). We observe that the entire OS level architecture is being shared across them. The only parts that are created from scratch are the *Bins* and *Libs*. Despite the process isolation and lightweight character, containers are less secure and more vulnerable compared to hypervisor-based virtualization, and thus, they can be used as alternatives of hypervisor-based virtualization.

Determining the most suitable technology for a specific scenario requires a thorough analysis of the scenario design and performance requirements, along with an ultra-careful analysis of the benefits and drawbacks associated with the use of one tool over the other. What can be observed is that virtualization tools can play a role towards energy efficiency (EE) improvement within MNs. Considering the hypervisor-based virtualization, the hypervisor can report resource usage to the orchestrator in order to trigger system automated sleep mode states and also to implement policies that are provided by management and orchestration entity, which includes power management and power stepping [24]. On the other hand, container-based virtualization can be considered as an alternative of hypervisor-based virtualization, as containers demand less memory space, portable and lightweight, and have a shorter start-up time which translates to low latency and power consumption [19].

1.1.2 NFV-based EPC architectural evolution

The works of [6, 7] proposed the use of a cloud-based approach with NFV platform, for the EPC evolution, in order to enable dynamic deployment of edge networks, the scaling of NFs, network monitoring, and load management. Both architectures avail the possibility of intelligently pooling capacity of resources when required.

In [6], the key elements of the architecture are (1) data-driven network intelligence for optimizing network resources usage and planning and (2) relaying and nesting techniques: to support multiple devices, group mobility, and nomadic hotspots. The EPC is virtualized into three parts, namely: (i) control plane entity (CPE), which is responsible for authentication, mobility management, radio resource control, and non-access stratum (NAS) and access stratum (AS) integration, (ii) the user plane entity (UPE), acting as a gateway, mobility anchor, and over-the-air (OTA) security provisioner. Lastly, (iii) the network intelligence (NI) plane is for the extraction of actionable insights from big data, orchestration or required services and functionalities (e.g., traffic optimization, caching, etc.). The realization of the network cloud can be achieved by enabling virtual function instances to be hosted in data centers when needed. The use of virtualization techniques will enable quick deployment and scalability of CPE and UPE functions. For example, in case of a natural disaster, with this technology, the local data center maybe unable to cope with the traffic upsurge; therefore, additional capacity can be sourced quickly from other data centers. A different strategy is employed in [7]. In that paper, the EPC architectural proposal simply abstracts the EPC network functions, decomposing and allowing them to run as software instances (virtual machines), on standard servers. This allows service providers to customize services and policies to design networks in new ways, to reduce costs, and simplify operations.

The aforementioned architecture proposals differ from one another. SDN is integrated with NFV in [6] to provide network control and to host the network intelligence, and in [7] only the NFV platform is available for enabling network services provision. Moreover, the architecture proposed in [7] is commercially available. Their contributions towards energy efficiency is as follows: (i) both make use of NFV and cloud computing platforms, and this avails the possibility of dynamically scaling resources based on demand as presented in [6] and (ii) through the information centric approach (collection of user-centric, network-centric, and context-centric data), intelligent algorithms, mainly network optimization tools, can be applied to the aggregated data in order to provide useful outlook for network planning and resource management.

2. Energy efficient in future mobile networks

There is growing awareness to the fact that the communication sector uses significant amount of energy [25]. This is especially true for wireless and in particular for the BSs of cellular networks, where energy costs make up a large part of the operating expenses of mobile operators. In addition, in future MNs, the BSs will be empowered with computation capabilities to enable local workload offloading (computation) and the provision of ultra-low latency services, as well as EH systems. This gives rise to the need for efficient energy management procedures employing machine learning and control-theoretic techniques, within the MEC paradigm.

2.1 Network infrastructure

As EH technologies advance, powering the network apparatuses (edge systems) with green energy (e.g., solar and/or wind) is a promising solution to reduce on-grid power due to their location, reliability, carbon footprint, and cost. The network infrastructure is illustrated in **Figure 2**, where a computing site is shown. The virtualized computing platform (MEC server) consists of the VMs (this can also be

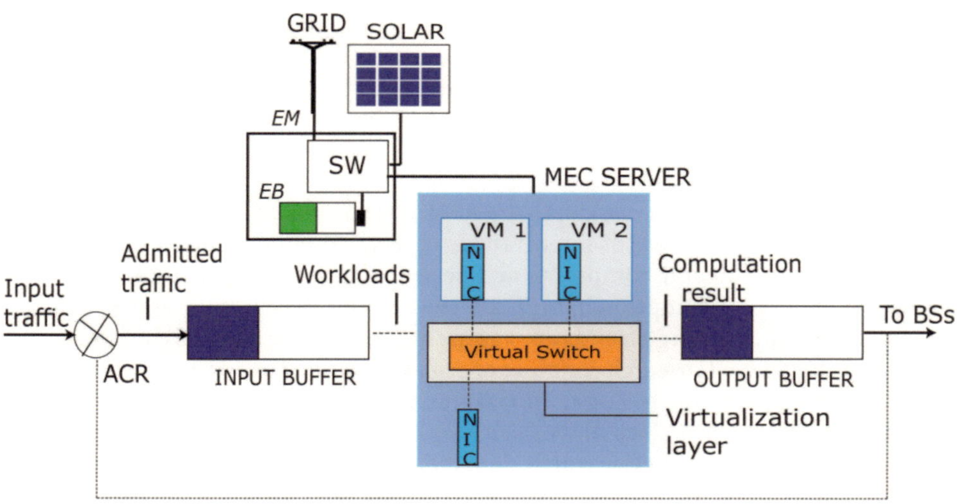

Figure 2.
The virtualized network infrastructure powered by hybrid energy sources: On-grid power and green energy. The electro-mechanical switch (SW) selects the appropriate source of energy [26].

containers), as the computing resources, empowered with EH capabilities through a solar panel and an energy buffer (EB) that enables energy storage. The power grid is also available for energy back up. The MEC node is assumed to be equipped with higher computational and storage resources compared to the end-user device. The energy manager (EM) is an entity responsible for selecting the appropriate energy source and for monitoring the energy level of the EB. The virtualized access control router (ACR) acts as an access gateway, responsible for routing, and it is locally hosted as an application. The input and output buffer are responsible for buffering admitted computation workload and storing aggregate computation results.

2.1.1 Energy consumption

As suggested by ETSI [12, 27], the virtualized network functions (VNFs) can be deployed at the base station (BS), that is, the BS site is empowered with computation capabilities where the MEC server is co-located with the BS or placed at an aggregation point (a point in close proximity to a group of BS) for edge network management.

We consider a MEC deployment scenario where the BS is co-located with a MEC server, as an example, and the total energy consumption ([J]) for the communication site is formulated as follows, inspired by [28, 29] and the virtualization knowledge from [21]:

$$\theta_{Tot}(t) = \theta_{BS}(t) + \theta_{MEC}(t),$$

where $\theta_{BS}(t)$ is the BS transmission energy (the load independent and the load dependent component), $\theta_{MEC}(t)$ is the energy drained due to computation and due to intra-communications processes in the MEC server at time slot t. It is worth noting that the computational (MEC) energy cost includes: (i) the energy drained due to the running computing resources (VMs or containers), w.r.t CPU utilization, (ii) the energy drained due to VM/containers switching their processing rates, and (iii) the energy induced by the transmission control protocol (TCP)/Internet protocol (IP) offload on the network interface card (NIC, partial computation at the network adapter such as TCP/IP checksum offload). Regarding the communication energy cost, we have energy drained due to the communication links (to-and-from each VM/container) and the energy drained due to the number of transmission (optical) drivers used for the data transfer to target BS(s).

2.2 Energy saving strategies in MEC

Considering the ultra-dense deployment of BSs and the dynamic characteristics of MNs, energy saving is of importance in green networks in order to exploit the benefits of sustainable computing and networking within edge systems. The sources of energy consumption in MEC are the BSs and the computing platforms. Towards this end, two large trends appear in the literature to address the energy efficiency challenge: (i) search for more energy efficient transmission devices and technologies; (ii) new technology proposals aimed at improving energy consumption in BSs, such as sleep mode, as it is known that switching on/off the BS transmission power during low traffic demand can yield significant energy savings [30–32]. Trend (ii) is more appealing due to the fact that future BS functions will be virtualized [1], thus enabling the ease of deactivating some of the BS functions. Within MEC servers, a joint *soft-scaling* (the reduction of computing resources per time instance) of the computing resources and the provision of minimum number of transmission drivers for data transfer to BS(s) in real-time is appealing.

2.2.1 Sleep-modes strategies in MNs

The dramatic growth in mobile data has spurred the dense deployment of small cell BSs to enhance spectrum efficiency and increase network capacity. Although small cell BSs consume less power compared with macro BSs, the overall power consumption of a large number of small cell BSs is phenomenal. To address the energy saving procedures, sleep mode is adopted. The sleep mode application can be observed from a single mobile operator or multiple operators perspective, with the main goal of maximizing energy savings without any significant network impact.

Towards energy savings in dense environment clustering algorithms have been proposed as a way of switching off BSs to reduce the energy consumption within MNs [33, 34]. However, the dynamic BS switching on/off strategies may have an *impact* on the network due to the traffic load that is offloaded to the neighboring BSs. To avoid this, the BS to be switched off must be carefully identified within a BS cluster. In [35], the *network impact* is used to identify the BS to be switched off within a cluster, one at a time, with no significant network performance degradation. Moreover, the network impact has been used in [36] to identify a BS to be switched off within a BS cluster where each BS is empowered with computation capabilities. With the advent of EH, it is desirable to incorporate the green energy utilization as a *performance metric* in traffic load balancing strategies [37, 38].

In the effort of greening future MNs, BSs are expected to be powered by green energy sources. On the other hand, NFV technology is expected to improve energy consumption by enabling the BS sleep modes, that is, scaling down the NFs during low traffic periods [24] or at low EB levels. Along the lines of MN softwarization, a distributed user association scheme that makes use of the soft-radio access network (RAN) concept for traffic load balancing via the RAN controller (RANC) is presented in [37]. Here, the user association algorithm is proposed and it runs in the RANC. The role of the algorithm is to enhance the network performance by reducing the average traffic delivery latency in BSs as well as to reduce the on-grid power consumption by optimizing the green energy usage. In cases where there is no centralized entity for edge network management, the need for distributed algorithm arises, which is the case with BSs co-located with MEC servers.

Furthermore, towards the effort of reducing the energy consumption through BS sleep modes, it is observed in the literature that most of the existing works considered clusters of BSs from a single mobile operator perspective, where some functions of the BS can be switched off and then the remaining active BSs handle the upcoming traffic. A new mechanism is presented in [39], which exploits the coexistence of multiple BSs from different mobile operators in the same area. An intra-cell roaming-based infrastructure-sharing strategy is proposed, followed by a distributed game-theoretic switching-off scheme that takes into account the conflicts and interaction among the different operators. Then, the work [40] of investigate the energy and cost efficiency of multiple HetNets [i.e., each HetNet is composed of eNodeBs (eNBs) and small cell BSs from one operator] that share their infrastructure and also are able to switch off part of it. Here, a form of roaming-based sharing is also adopted, whereby the operator can roam its traffic to a rival operator during a predefined period of time and area. An energy efficient optimization problem is formulated and solved using a cooperative greedy heuristic algorithm.

2.2.2 Energy savings in virtualized platforms using soft-scaling

To address 5G use cases in a more energy efficient way, visibility into power usage is required for developing power management policies in virtualized computing platforms (i.e., MEC servers or data centers). In virtualized computing platforms, the energy consumption is related to the computing and communication processes. To minimize energy consumption within such environments, energy saving studies have involved the scaling up/down of servers/VMs [29, 41–43], VM migration (process of duplicating and transmitting the VM memory image over the network, in virtualized data centers, without or least service interruption [44]), and soft resource scaling (shortening the access time to physical resources [45]). With the advent of NFV, it is expected that the NFV framework [24, 46] can exploit the benefits of virtualization technologies to significantly reduce the energy consumption of large scale network infrastructures. In addition, the EE control framework [47] shall provide the control sequence and procedures for controlling and managing energy efficiency, within self-managed automated edge systems. Virtualization provides a promising approach for consolidating multiple online services onto few computing resources within an enterprise data center. By dynamically provisioning VMs, consolidating the workload, and switching servers on/off as needed, service providers can maintain the desired QoS while achieving higher server utilization and EE. In [17], a dynamic resource provisioning framework for a virtualized computing environment is presented and it is experimentally validated on a small server cluster that provides online services. Along the lines of MEC [28, 36], the long-term short memory (LSTM) neural network is used for forecasting the traffic load and harvested energy, and then the limited lookahead control (LLC) technique uses the forecasted traffic load and harvested energy to obtain the best control input that will drive the edge systems into the desired behavior.

3. Mobile datasets for network solutions

To enable power transmission adaptation and load balancing across BSs, traffic load information (operator mobile traffic datasets) obtained from the EPC can be utilized to extract relevant demand patterns to design dynamic BS management mechanisms [48, 49]. Optimization based on mobile traffic datasets, obtained from multiple network elements described in [49], will make it possible to minimize the amount of time it takes to steer traffic on a real time basis, thus provisioning the network resources for computing and communication. From a networking perspective, the understanding and characterization of the traffic consumption within the network can pave the way towards more efficient and user/traffic-oriented networking solutions.

A greedy algorithm is used to estimate the energy savings through dynamic BS operation based on real cellular traffic traces and actual BS location, within an urban environment [48]. In addition, during crowded events (e.g., concerts and soccer games), MNs face voice and data traffic volumes that are often orders of magnitude higher than what they face during normal days. In [50], datasets usage show the potential of dynamically switching on/off BSs around a stadium (soccer field), as some of the BSs experience low traffic load during a large event. Moreover, it is shown that the use of real-world voice and data traces can provide insights about when to tune the radio resource allocation and to make use of opportunistic connection sharing (the aggregating of traffic from multiple devices into a single cellular connection) towards network improvement, without incurring any costs related to infrastructure changes [51]. To improve energy savings and guarantee QoS, within the MEC paradigm, in [28, 36] the traffic load and energy datasets (solar and wind) are used to provide a short-term forecast, that is then used by the foresighted optimization algorithm towards dynamic resource allocation.

4. Conclusion

Future mobile networks are envisioned to be softwarized and the network functions virtualized. In light of this evolution, MEC has emerged to enable ultra-low latency services, distributed intelligence, and location-aware data processing at the network edge. The foreseen dense deployment of base stations empowered with computation capabilities motivates the need for energizing the edge systems (BSs and MEC servers) with green energy (solar and/or wind energy) in order to minimize the carbon footprint and the dependence on the power grid. Moreover, the use of green energy promotes energy self-sustainability within the network. In order to improve the energy efficiency, mobile traffic datasets can be used to come up with traffic-oriented network management solutions. For dynamic resource management, over a look-ahead prediction horizon, machine learning methods and control-theoretic techniques (foresighted optimization) are foreseen as tools for edge network management, i.e., for green energy-aware network apparatuses.

Author details

Thembelihle Dlamini
University of Padova, Padova, Italy

*Address all correspondence to: dlamini@dei.unipd.it

IntechOpen

© 2019 The Author(s). Licensee IntechOpen. This chapter is distributed under the terms of the Creative Commons Attribution License (http://creativecommons.org/licenses/by/3.0), which permits unrestricted use, distribution, and reproduction in any medium, provided the original work is properly cited.

References

[1] Virtualization for small cells: Overview. Small Cell Forum, Draycott, England, Tech. Rep.; 2015

[2] Thembelihle D, Rossi M, Munaretto D. Softwarization of mobile network functions towards agile and energy efficient 5G architectures: A survey. Wireless Communications and Mobile Computing. 2017;**2017**:21. Article ID: 8618364. DOI: 10.1155/2017/8618364

[3] Xin J, Li LE, Laurent V, Rexford J. Softcell: Scalable and flexible cellular core network architecture. In: ACM Conference on Emerging Networking Experiments and Technologies. Santa Barbara, USA; 2013

[4] Hawilo H, Shami A, Mirahmadi M, Asal R. NFV: State of the art, challenges, and implementation in next generation mobile networks (vEPC). IEEE Network. 2014;**28**(6):18-26. DOI: 10.1109/MNET.2014.6963800

[5] Yang C, Chen Z, Xia B, Wang J. When ICN meets C-RAN for HetNets: An SDN approach. IEEE Communications Magazine. 2015;**53**(11):118-125. DOI: 10.1109/MCOM.2015.7321980

[6] Agyapong PK, Iwamura M, Staehle D, Kiess W, Benjebbour A. Design considerations for a 5G network architecture. IEEE Communications Magazine. 2014;**52**(11):65-75. DOI: 10.1109/MCOM.2014.6957145

[7] Unlock New Business Opportunities With vEPC. VMware, USA, Tech. Rep.; 2016

[8] 5G Mobile Communications Systems for 2020 and beyond. Fifth generation mobile promotion forum, Japan, Tech. Rep.; 2016

[9] Nakao A, Ping D, Masayuki K. Flare: Deeply programmable network node architecture. EU-Japan Collaboration Project, University of Tokyo, Japan; 2016. [Online]. Available from: http://www.5gsummit.org/berlin/docs/slides/Aki-Nakao.pdf

[10] M-CORD Open Reference Solution Paves the Way for 5G Innovation. 2016. [Online]. Available from: http://opencord.org/tag/m-cord/

[11] Nakao A. Software-defined data plane enhancing SDN and NFV. IEICE Transactions on Communications. 2015;**98**(1):12-19. DOI: 10.1587/transcom.E98.B.12

[12] Kekki S, Featherstone W, Fang Y, Kuure P, Li A, Ranjan A et al. MEC in 5G Networks. ETSI, Sophia-Antipolis, France, Tech. Rep.; 2018

[13] Patel M, Hu Y, Hédé P, Joubert J, Thornton C, Naughton B et al. Mobile edge computing introductory technical white paper. ETSI, Sophia-Antipolis, France, Tech. Rep.; 2014

[14] Lu G, Guo C, Li Y, Zhou Z, Yuan T, Wu H et al. ServerSwitch: A programmable and high performance platform for data center networks. In: USENIX Conference on Networked Systems Design and Implementation. Boston, USA; 2011

[15] Nunes BAA, Mendonca M, Nguyen XN, Obraczka K, Turletti T. A survey of software-defined networking: Past, present, and future of programmable networks. IEEE Communication Surveys and Tutorials. 2014;**16**(3):1617-1634. DOI: 10.1109/SURV.2014.012214.00180

[16] Ciosi M, Clarke D, Cui C, Benitez J, Mivhel U, Ogaki K et al. Introductory white paper: Network functions virtualization. In: SDN and Open Flow World Congress. Darmstadt, Germany; 2012

[17] Kusic D, Kephart JO, Hanson JE, Kandasamy N, Jiang G. Power and performance management of virtualized computing environments via lookahead control. In: International Conference on Autonomic Computing, Chicago, USA; 2008

[18] Bhushan N, Li J, Malladi D, Gilmore R, Brenner D, Damnjanovic A, et al. Network densification: The dominant theme for wireless evolution into 5G. IEEE Communications Magazine. 2014;**52**(2):82-89. DOI: 10.1109/MCOM.2014.6736747

[19] Morabito R. Power consumption of virtualization technologies: An empirical investigation. In: IEEE International Conference on Utility and Cloud Computing (UCC). Limassol, Cyprus; 2015

[20] Jin Y, Wen Y, Chen Q. Energy efficiency and server virtualization in data centers: An empirical investigation. In: IEEE Conference on Computer Communications Workshops (INFOCOM Workshops). Orlando, USA; 2012

[21] Portnoy M. Virtualization Essentials. Indianapolis, Indiana, USA: John Wiley & Sons; 2012

[22] Wu B, Fu S, Jiang X, Wen H. Joint scheduling and routing for QoS guaranteed packet transmission in energy efficient reconfigurable WDM mesh networks. IEEE Journal on Selected Areas in Communications. 2014;**32**(8):1533-1541. DOI: 10.1109/JSAC.2014.2335313

[23] Fu S, Wen H, Wu J, Wu B. Cross-networks energy efficiency trade-off: From wired networks to wireless networks. IEEE Access. 2017;**5**:15-26. DOI: 10.1109/ACCESS.2016.2585221

[24] Network Functions Virtualisation (NFV): Hypervisor Domain. ETSI, Sophia-Antipolis, France, Tech. Rep.; 2015

[25] Cisco visual networking index: Forecast and Methodology, 2017-2022. Cisco, Tech. Rep.; 2019

[26] Dlamini T, Gambin AF. Adaptive resource management for a virtualized computing platform in edge computing. In: IEEE International Conference on Sensing, Communication and Networking (SECON). Boston, USA; 2019

[27] Gabriel B. Mobile edge computing use cases and deployment options. Heavy Reading, Tech. Rep.; 2016

[28] Dlamini T, Gambín ÁF, Munaretto D, Rossi M. Online resource management in energy harvesting BS sites through prediction and soft-scaling of computing resources. In: IEEE PIMRC. Bologna, Italy; 2018

[29] Shojafar M, Cordeschi N, Amendola D, Baccarelli E. Energy-saving adaptive computing and traffic engineering for real-time-service data centers. In: IEEE International Conference on Communication Workshop (ICCW), London, UK; 2015

[30] Bousia A, Kartsakli E, Antonopoulos A, Alonso L, Verikoukis C. Multi-objective auction-based switching-off scheme in heterogeneous networks: To bid or not to bid? IEEE Transactions on Vehicular Technology. 2016;**65**(11):9168-9180. DOI: 10.1109/TVT.2016.2517698

[31] Dongsheng Han BZ, Chen Z. Sleep mechanism of base station based on minimum energy cost. Wireless Communications and Mobile Computing. 2018;**2018**:13. Article ID: 4202748. DOI: 10.1155/2018/4202748

[32] Zhu Y, Zeng Z, Zhang T, An L, Xiao L. An energy efficient user association scheme based on cell sleeping in LTE heterogeneous networks. In: International Symposium on Wireless Personal Multimedia

Communications (WPMC). Sydney, Australia; 2014

[33] Zhang H, Cai J, Li X. Energy-efficient base station control with dynamic clustering in cellular network. In: IEEE International Conference on Communications and Networking (CHINACOM). Guilin, China; 2013

[34] Samarakoon S, Bennis M, Saad W, Latva-aho M. Dynamic clustering and ON/OFF strategies for wireless small cell networks. IEEE Transactions on Wireless Communications. 2016;**15**(3):2164-2178. DOI: 10.1109/TWC.2015.2499182

[35] Oh E, Son K, Krishnamachari B. Dynamic base station switching-on/off strategies for green cellular networks. IEEE Transactions on Wireless Communications. 2013;**12**(5):2126-2136. DOI: 10.1109/TWC.2013.032013.120494

[36] Dlamini T, Gambín ÁF, Munaretto D, Rossi M. Online supervisory control and resource management for energy harvesting BS sites empowered with computation capabilities. Wireless Communications and Mobile Computing. 2019;**2019**:17. Article ID: 8593808. DOI: 10.1155/2019/8593808

[37] Han T, Ansari N. A traffic load balancing framework for software-defined radio access networks powered by hybrid energy sources. IEEE/ACM Transactions on Networking. 2016;**24**(2):1038-1051. DOI: 10.1109/TNET.2015.2404576

[38] Xu J, Wu H, Chen L, Shen C, Wen W. Online geographical load balancing for mobile edge computing with energy harvesting. arXiv preprint arXiv:1704.00107; 2017

[39] Antonopoulos A, Kartsakli E, Bousia A, Alonso L, Verikoukis C. Energy-efficient infrastructure sharing in multi-operator mobile networks. IEEE Communications Magazine. 2015;(5):242-249. DOI: 10.1109/MCOM.2015.7105671

[40] Oikonomakou M, Antonopoulos A, Alonso L, Verikoukis C. Evaluating cost allocation imposed by cooperative switching off in multi-operator shared HetNets. IEEE Transactions on Vehicular Technology. 2017;**66**(12):11352-11365

[41] Xu J, Ren S. Online learning for offloading and autoscaling in renewable-powered mobile edge computing. In: IEEE Global Communications Conference (GLOBECOM). Washington, USA; 2016

[42] Shojafar M, Cordeschi N, Baccarelli E. Energy-efficient adaptive resource management for real-time vehicular cloud services. IEEE Transactions on Cloud Computing. 2016;7(1):196-209. DOI: 10.1109/TCC.2016.2551747

[43] Guenter B, Jain N, Williams C. Managing cost, performance, and reliability tradeoffs for energy-aware server provisioning. In: Proceedings IEEE INFOCOM. Shanghai, China; 2011

[44] Beloglazov A, Abawajy J, Buyya R. Energy-aware resource allocation heuristics for efficient management of data centers for cloud computing. Future Generation Computer Systems. 2012;**28**(5):755-768

[45] Nathuji R, Schwan K. VirtualPower: coordinated power management in virtualized enterprise systems. In: ACM SIGOPS symposium on Operating systems principles. Washington, USA; 2007

[46] ETSI GS MEC: Mobile Edge Computing (MEC); Framework and Reference Architecture. ETSI, France, Tech. Rep.; 2016

[47] Study on Energy Efficiency Aspects of 3GPP (3GPP TR 21.866 - Release 14). ETSI, France, Tech. Rep.; 2016

[48] Oh E, Krishnamachari B, Liu X, Niu Z. Toward dynamic energy-efficient operation of cellular network infrastructure. IEEE Communications Magazine. 2011;**49**(6). DOI: 10.1109/MCOM.2011.5783985

[49] Charging management; Charging Data Record (CDR) parameter description (3GPP TS 32.2 .298 version 13.5.0 Release 13). ETSI, France, Tech. Rep.; 2016

[50] Erman J, Ramakrishnan KK. Understanding the super-sized traffic of the super bowl. In: Proceedings of the 2013 conference on Internet measurement conference. Barcelona, Spain; 2013

[51] Zubair SM, Lusheng J, X LA, Jeffrey P, Shobha V, Jia W. A first look at cellular network performance during crowded events. ACM Sigmetrics Performance Evaluation Review. 2013;**41**(1):17-28

Chapter 5

Localization Enhanced Mobile Networks

Salman Al-Shehri, Pavel Loskot and Michael J. Hirsch

Abstract

The interest in mobile ad-hoc networks (MANETs) and often more precisely vehicular ad-hoc networks (VANETs) is steadily growing with many new applications, and even anticipated support in the emerging 5G networks. Particularly in outdoor scenarios, there are different mechanisms to make the mobile nodes aware of their geographical location at all times. The location information can be utilized at different layers of the protocol stack to enhance communication services in the network. Specifically, geographical routing can facilitate route management with smaller overhead than the traditional proactive and reactive routing protocols. In order to achieve similar advantages for radio resource management (RRM) and multiple access protocols, the concept of virtual cells is devised to exploit fully distributed knowledge of node locations. The virtual cells define clusters of MANET nodes assuming a predefined set of geographically distributed anchor points. It enables fast response of the network to changes in the nodes spatial configuration. More importantly, the notion of geographical location can be generalized to other shared contexts which can be learned or otherwise acquired by the network nodes. The strategy of enhancing communication services by shared contexts is likely to be one of the key features in the beyond-5G networks.

Keywords: context, distributed protocol, localization, MANET, radio resource management, routing, VANET

1. Introduction

The support for mobility was a large step towards realizing the full potential of wireless networks. The mobility of nodes brings about two large concerns. It affects radio propagation conditions making the propagation channels between transmitters and receivers more volatile and less predictable. It also complicates the network management at higher layers of the protocol stack, since the network need to be aware about the present locations of all mobile nodes. The solutions to address these two concerns are fundamentally dependent whether there is a supporting infrastructure such as fixed base stations and access points, or whether the nodes can only communicate directly with each other. The former scenario was introduced with the first generations of cellular networks whereas the latter scenario appeared in MANETs. The emerging 5G networks are expected to provide support not only to individual mobile nodes, but newly also to MANETs. Alternative strategy to conventional networks comprising named nodes are the so-called data-centric networks which were also assumed for MANETs. In these networks, the nodes advertise and replicate named data, so the network routing is driven by the requests for given data rather than for given nodes.

 IntechOpen

Most MANETs are formed by interconnected manned or unmanned vehicles on the ground or in the air, so they are also referred to as VANETs. Many new applications are envisioned particularly for networks of unmanned aerial vehicles (UAVs) or drones, and other high altitude platforms (HAPs) such as balloons [1]. Another prominent example of MANETs are the upcoming networks of the low-Earth orbit (LEO) satellite networks where inter-satellite communications will be a critical component for delivering the envisioned broadband services and the global Earth coverage.

The radio propagation environment and the node mobility drive intermittent and often unpredictable connectivity between nodes in MANETs. The challenge is to define the corresponding mathematical models which are tractable as well as sufficiently accurate [2]. At minimum, the radio propagation models need to incorporate path-loss, random shadowing and multi-path shadowing can be approximated by a two-ray ground reflection model. The mobility models require much more sophisticated strategies to account for individual and group behaviors of nodes including responses to various events, terrain profile, physical laws and many other aspects [3]. The discrepancy between the measurements in real-world networks and the protocol performances predicted from simulations can be largely attributed to inaccurate or inappropriate mobility models.

The dynamic nature of MANETs necessitates development of bespoke protocols, since conventional protocols such as TCP/IP used in wired networks would be very inefficient or even unusable, mainly due to very large overhead. For instance, MANETs require frequent packet retransmissions, re-establishing network routes to maintain connected paths between nodes, session management to deal with dropped connections, and security provisioning against internal and external attacks. Moreover, the bandwidth and packet payload is often limited, and the nodes may have reduced computing, communication and storage capabilities. This calls for carefully designed protocols to optimize the resource, so it is not surprising that protocol suits in the commercial MANETs are often proprietary, possibly modified versions of the protocols from research literature. Practical implementation of protocols also faces many common issues of software development including hidden bugs which may be extremely difficult to discover.

At the physical layer, the node mobility creates fast fading channels which can be mitigated by various diversity signaling techniques including error-correction channel coding schemes, multicarrier modulations, and multiple antennas systems. In MANETs, the mobility is limited to a given geographical area, and the nodes participating in the network are usually known beforehand. This simplifies the protocols for mobility management in MANETs by allowing fixed node identifiers. On the other hand, MANETs are more vulnerable to security attacks than cellular networks. For example, there is no centralized authority in MANETs which can be trusted, and relatively short lifespan and small traffic volumes do not allow statistically significant intrusion detection.

The main strategy of the upcoming 5G networks is to unite telecommunication systems and provide unified and transparent access in different scenarios using different technologies. Hence, the 5G networks should provide support for MANETs as well. However, unlike (D2D) single-hop communication links in the long-term evolution (LTE) 4G networks, the MANET support in the 5G networks is likely to enable more flexible integration of mobile sub-networks within the cellular infrastructure with computing centers. Especially the VANETs of connected vehicles on the ground or in the air is a highly anticipated application supported in the 5G systems. However, some degree of autonomy required for MANETs or VANETs while exploiting the 5G infrastructure if or when it is available will make the orchestration of communication and computing resources in these networks extremely challenging. Exploiting the location information of mobile nodes could significantly reduce the complexities of network control and management in the envisioned 5G systems.

Several key network services which must be provided in mobile networks are discussed in Section 2. We describe mobility management, and introduce different types of context. Geographical location is shown to be a specific case of a shared context within telecommunication networks which can be utilized to enhance the network services. We also briefly outline localization services in the 4G and 5G networks, since these networks are expected to support MANETs in future. In Section 3, we review conventional and geographical routing strategies in mobile networks that have been studied extensively in literature. In contrast, geographical RRM and multiple access schemes received much less attention in the literature. A new concept of virtual cells for geographical protocols at the link layer providing a fast response with minimum overhead to varying MANET topology is presented in Section 4. The chapter is concluded in Section 5.

2. Network services in mobile networks

2.1 Mobility management

We review three concepts which are crucial for decentralized applications in MANETs: mobility management, network contexts, and localization services. In particular, the applications in MANETs need to be at least partially distributed including node localization. The distributed applications rely on and are greatly affected by the characteristics of inter-node connectivity such as time varying capacity of links. The end-to-end path stability and delay is also affected by the network traffic load with possible congestion effects, the number of hops and the number of alternative routes between the source and the sink. From computing perspective, the mobility management requires information about locations of clients and server instances, and maintaining states of sessions to provide robustness against disruptions. The applications offering suspend and resume functions are less common in highly dynamic MANETs. Provided that there is enough bandwidth and additional latency can be tolerated, off-loading applications into a cloud solves the computing constraints of nodes. Distributed clouds known as cloudlets which are more easily accessible by the network nodes were introduce to balance the requirements for bandwidth, latency and computing. The resource utilization is optimized by profiling applications, devices and network connectivity. The recent trend is to run synchronized identical instances of an application in the network nodes as well as in the cloud in order to optimize fine-grain off-loading in real-time [4]. The application can call micro-services to alleviate the latency for setting up and configuring full virtual machine instances while utilizing more efficiently the cloud resources.

In highly mobile environments such as in MANETs, the decentralized applications can be implemented as smart messages combining data and code [4]. The code manipulating data is executed as needed along the route as the message is passed among the nodes. This approach offers good scalability while executing the application within a desired context, for example, when the message reaches a node in a given location. Smart messages also solve the problem of migrating services among nodes as needed. In addition, akin to data-centric networks with named data chunks, it is possible to use smart messages with unique global names to be requested by the network nodes.

In the 5G networks, the nodes in MANETs can benefit from mobility management mechanisms including tracking area lists and NAS (non-access stratum) messages, provided that these nodes are governed directly or indirectly by the 5G network controllers. The interesting and open research question is how to manage the mobility in networks where some but not all nodes in a MANET are controlled by the 5G network. Another open research question is how to exploit predicted node trajectories to simplify the mobility management by inferring the future node positions.

2.2 Context in mobile networks

Context in telecommunication networks can have different meanings [5]. It can be related to some objective in delivering telecommunication services which is supported by directly observable or implied conditions. Typical characteristics describing the context in telecommunication networks are following:

- Context can be defined locally or globally, and it can be set, managed, synchronized, combined, and transferred.
- Context often varies in time, but context-based adaptation can go beyond simply adjusting a few parameters, for example, to improve efficiency and resilience of the network.
- Context usually describes more complex conditions in the network, and it can be defined for a single node, a group of nodes or all nodes in network.
- Context can be shared, and learned individually or cooperatively, or predicted from past observations.

Context sharing is illustrated in **Figure 1** where either information about individual contexts is shared explicitly, or some shared network conditions are observed individually by different network entities. In some cases, revealing the context such as geographical location could cause privacy and security concerns, which may require defining and enforcing context sharing policies. A trivial example of the shared context is time synchronization of nodes in a network required, for example, to define time slots for multiple access protocol.

We can assume different types of contexts such as context defined for connectivity, devices, applications and networks including availability of different resources. Here, our focus is specifically on geographical location as the context which is naturally shared among the network nodes. Similarly to time context, the shared location context can be acquired with the aid of an external source such as

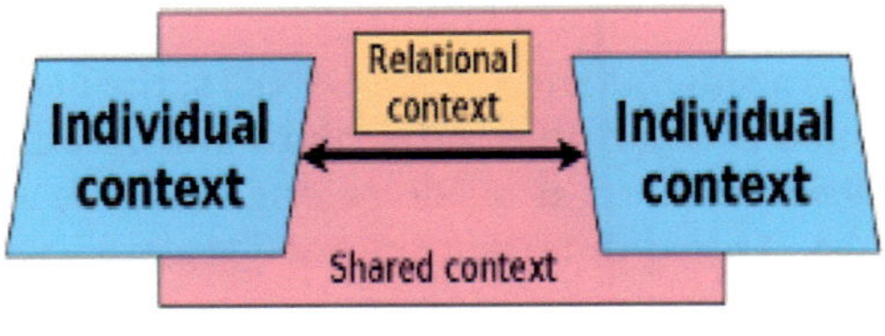

Figure 1.
Sharing the context among different network entities.

global satellite navigation system (GNSS), or the nodes can cooperate to define their locations relative to each other. Moreover geographical locations can be defined in the same or multiple spatial frames with the corresponding points of origin. In addition, geographical location can be defined more loosely as a position belonging to some specified geographical area such as a base station cell, inside the building and similar. Such coarse-grained localization is often sufficient in many applications, for example, to make off-loading decisions.

2.3 Localization in mobile networks

The most straightforward for determining the absolute locations of nodes in a geographical area is to employ GNSS which is cost-affordable technology with ubiquitous coverage outdoors. Recently, a number of countries launched their own now fully operational GNSS including USA (GPS), GLONASS (Russia), Galileo (Europe), Compass (China), and IRNSS (India). Localization errors of GNSS can be improved by correcting errors due to atmospheric propagation effects, using overlay signals from other satellites, and using terrestrial augmentation systems. Another strategy particularly suitable for mobile nodes is to employ inertial navigation systems (INS) onboard the nodes to perform dead reckoning. The INS can be used as a fallback system when the GNSS signals are temporarily unavailable, for instance, in between the satellite measurements.

Localization techniques in mobile networks which are independent of external signals assume measurements of signal strength, time of light, time differences, angle of arrival and others [6]. However, the measurements are always noisy, so more sophisticated statistical signal processing such as Kalman filtering is usually necessary. The localization by inferring distances to several other nodes known as trilateration is probably the most common. There are also network assisted localization methods which will support mobile networks in future (5G) systems. For instance, the node may inquiry about the identifier of the base station, or it can identify some other suitable radio beacon nearby to determine its approximate location. The 4G/5G base stations can assist the GNSS localization in order to reduce the location acquisition time. The 4G/5G standard also defines several references signals to assist the mobile nodes in measuring signal strength and observed difference in time of arrival for determining their location.

The positioning methods defined in the latest LTE standard adopted as the New-Radio (NR) 5G system are intended to provide a broad compatibility with other radio access technologies and exploit different measurements, especially in the uplink. The LTE positioning protocol (LPP) can assist the mobile nodes in determining their location using the control plane or user plane signaling. For instance, the base station can calculate the position using the GNSS measurements reported by the node, or the base station can provide the current satellite data to the node to facilitate its GNSS positioning. However, in situations when the GNSS signal is not be available, the preferred localization method in the 4G networks is based on the observed time difference of arrival (OTDOA) at the node from two or more base stations. The timing advance information which is used to synchronize multiple base station cells can be used for node localization. The locations of base stations are exactly known, so they can be utilized as anchors in conventional wireless localization techniques. There are also specifically defined positioning reference signals for signal timing and strength measurements, and the radio-frequency signature inference in the LTE. The open research question is how to provide network assisted localization services for MANETs where only some nodes are controlled by the 5G network.

3. Routing in mobile networks

3.1 Conventional routing protocols

Routing is a primary function of the network layer. The routing strategy is one of the key factors affecting the achievable QoS in the network. It is usually a compromise between fairness and traffic prioritization. The routing protocols need to define services for route creation, maintenance, updates, release and deletion, and it can also provide backup path to a faster recovery from link failures. The routing protocols in MANETs are fully distributed, and need to support mobility and the dynamic network topology, so they are often some variation of adaptive distance vector routing. The lifespan of links and routes in MANETs is dependent on the node mobility. The routing algorithm needs to be robust in order to improve the network stability despite existence of short-lived links. The local coordination among nodes is necessary to create longer-lived routes. The neighboring nodes periodically and iteratively share their knowledge of the dynamic network topology. The neighboring nodes are commonly discovered by sending hello and echo messages, for instance, using selective or controlled flooding. A simple flooding suffers from packet duplication to the point of packet implosion due to routes overlap. A straightforward improvement of flooding known as gossiping assumes a random walk to forward packets to the randomly selected outgoing links, but it cannot guarantee that all packets will reach all destinations. However, any routing protocols based on flooding do not scale well with the network size, if the network topology remains flat.

Apart from flooding, other basic mechanisms for route discovery assume next hop routing, and source based routing. The hello messages are also used to probe existing connections, if there were no other packets sent within a given time period to ensure that the neighboring nodes are still available.

The fundamental requirement for routing protocols is to discover optimum or near optimum routes which are loop-free. The loop-free routing is related to count-to-infinity problem which can occur if one of the intermediate routers goes down, or the routing updates between two or more nodes appear at the same time. The routing path optimality can be measured as end-to-end latency and bandwidth which can be approximated by the number of hops, or geographical distances. In many scenarios, the optimum routing is constrained by availability and fair use of resources. Although QoS-aware routing in MANETs is less common, the energy-aware routing algorithms are frequently assumed to avoid exhausting the battery life of the nodes serving as routers for all the other nodes. This can be achieved by periodically changing the group of nodes assigned to act as routers. The energy dissipation in nodes is greatly affected by the uniformity of traffic in the network. The battery life can be also extended by defining duty cycles with sleep modes and periodic awakening.

The routing is often combined with scheduling which can be reservation based to avoid collisions, or contention based scheduling is more efficient with smaller network traffic loads. The routing defines a particular network topology such as a chain topology which is useful for data aggregation, and cycle-free spanning tree for packet broadcasting and multicasting. Determining spanning tree is, however, problematic in dynamic networks where it is usually approximated, for example, using a reverse-path forwarding mechanism. The spanning tree topology can be also established at the level of multicast groups, and there can be multiple spanning trees from the same source to different multicast groups.

The data aggregation creates ever larger payload as the packet traverse along the route in exchange of reducing the number of packets to be sent. The overhead of routing protocols increases substantially with the network size and its dynamics.

The updates via control messages consume the bandwidth and energy. The frequency of updates determines the temporal resolution, i.e., the maximum dynamics of network which can be supported. It is also possible to limit the spatial resolution of updates by constraining how far they can propagate in the network. This issues are more problematic for flat peer-to-peer MANETs, so creating a two-tier hierarchy of nodes by assigning nodes to clusters is usually desirable. The clusters are created by clustering algorithms, and each cluster elects a cluster head to forward packets to other clusters whereas the nodes in the cluster can communicate directly. The clustering of network reduces the number of hops to destination which reduces the end-to-end delay. The geographical distances between clusters can be measured by assuming the cluster centroids.

In general, different routing protocols are required for different scenarios and applications. The basic classification of routing protocols whether they provide route discovery on demand or a priori. These two classes are referred to as reactive and proactive protocols, respectively. Reactive protocols start the route discovery only when it is needed, i.e., there are data to be transported over the network. The process is initiated by the source with the data which avoids the need for routing tables in intermediate nodes and their periodic updates. However, the routing overhead and the packet payload increases with the number of hops as the route grows towards the destination, and each intermediate node appends its identifier to the packet header. However, the large packet size can create problems for the link layer protocol as it normally sends packets of predefined length, and larger packets must be sliced into multiple pieces. The route discovery is supported by broadcasting RREQ (route request) and receiving RREP (route reply) messages.

The discovered routes can be cached to improve efficiency and reduce the control overhead. However, the cached routing information eventually becomes stale. In addition, sudden changes to the route such as broken links are not detected. The on-demand routing strategy is more efficient for less frequent data transfers, and when the network topology is less dynamic, even though there is some delay before the route is set up. The reactive protocols are usually based on distance vector routing. The most well-known examples of reactive routing protocols in MANETs are AODV, DSR, and TORA [7]. For instance, the DSR protocol is useful for unicast traffic with multiple routes between source and destination, but it suffers from the growing packet size. The AODV protocol has constant packet sizes by keeping routing information in routing tables at intermittent nodes. Each route is assigned an expiry time, and only the routes in use are maintained. In addition, the sequence numbers in packets are used to keep track of active routes. The AODV protocol also supports multicast routing.

Next-hop routing protocols optimize only the following hop unlike the source based routing which considers the whole end-to-end path to the destination. The reactive protocols can update the route if the detected changes are above a certain threshold in order to reduce the frequent route updates in time-sensitive applications. The TEEN protocol is an example of this approach. The diffusion routing protocols propagate data along the reverse path of the initial query. Each path is associated with a gradient which is formed by propagating the initial data query or so-called the interest message. The data-query based routing protocols are unsurprisingly used in data-centric networks. The SPIN protocol is one example of these kinds of protocols.

Alternative strategy to reactive protocols is to assume proactive protocols which establish routes a priori, even if there are no data to be sent. This routing strategy is more suited to networks which larger traffic loads, but smaller mobility compared to reactive protocols. The proactive protocols can assume both distance vector and link state routing algorithms, and since they primarily rely on routing tables. The routing protocols can afford to search for the shortest path or the least cost route, and exploit multiple routes between the source and destination. The timers as well as sequence numbers are again utilized to detect stale routes and remove them from routing tables. The main disadvantage is periodic dissemination of routing information to maintain the routing tables.

The LEACH protocol is a popular example of proactive protocol used with topology clustering. The PEGASIS protocol improves the LEACH protocol, and uses sequential data aggregation over chain topology, although parallel aggregation strategies were also considered. Both these protocols are much more efficient for broadcasting than flooding based algorithms, since they assume topology clustering, however, their support for mobility is limited, and there are no considerations for QoS provisioning. Other examples of proactive protocols include OLSR, DSDV and WRP.

There are also hybrid routing protocols combing reactive and proactive protocols to maximize the benefits of both, for example, GRP and ZRP protocols.

3.2 Geographical routing protocols

Unlike previously described topology-based routing protocols, the geographical protocols derive their routing strategy from knowledge of geographical locations of other nodes in the network [7–12]. The geographical locations can be also used to forecast the propagation conditions either by simple mathematical models if it needs to be performed in real time or by simulations if the off-line channel modeling is acceptable. Geographical locations can also represent the network node addresses, but this is less practical in mobile networks. There is also a concept of so-called virtual embeddings which assign the nodes with virtual stationary points serving as their addresses in order to alleviate the need for determining the actual geographical positions of each node.

The key idea of geographical routing which goes back to 1980 is to forward packets closer to the destination without prior path discovery, similarly to reactive routing protocols. Hence, geographical routing is particularly useful for MANETs with frequent topology changes, provided that the geographical locations of the neighboring nodes can be tracked. The common challenges of geographical routing protocols are difficulty in obtaining geographical locations of other nodes apart from the immediate neighbors, and accuracy and timeliness of determining the locations. Another issues is timeliness of location information as the nodes in MANETs are constantly moving, and before the relevant information is forwarded to other nodes, it may be obsolete. The performance of these protocols can be improved by predicting node locations knowing their mobility patterns which can be then used to predict the quality of links. It should be also noted that these protocols were primarily developed for 2D locations. The extension to 3D space including the airborne nodes may not be straightforward.

There are two basic strategies employed in geographical routing protocols. The first strategy is the one-hop greedy forwarding. The idea is to bring packet closer to the destination. As illustrative example in **Figure 2**, the source is connected to four nodes A, B, C, and D within its transmission radius. The node A is selected as the nearest node providing a forwarding progress towards the destination. The node B offers the best forwarding progress towards the destination among all the nodes connected with the source. The node C is selected as the one being closest to the azimuth towards the destination, so this strategy is referred to as compass routing. Finally, selecting the node D as the one being closest to the destination corresponds to a basic greedy strategy. More importantly, neither compass routing nor the nearest node with forwarding progress guarantees the loop free routing. The greedy forwarding can lead to a dead-end once there are no other nodes closer to the destination.

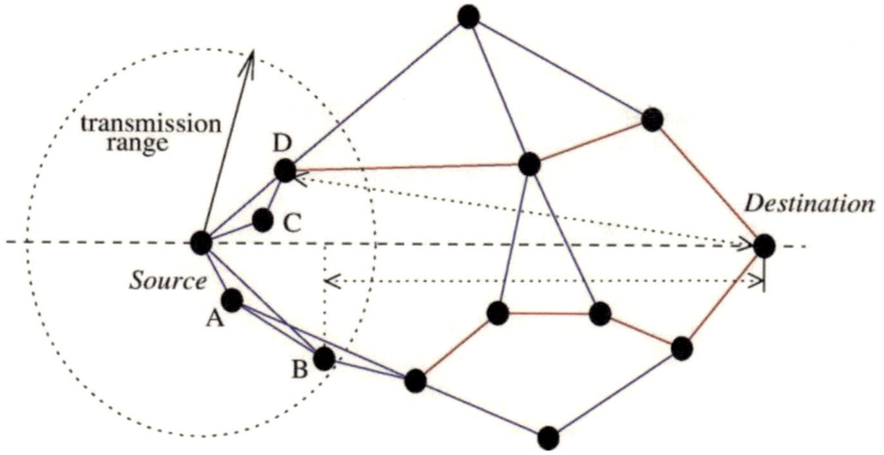

Figure 2.
Greedy one-hop forwarding and face routing in geographical routing protocols.

The second strategy is known as face routing. The faces are polygons depicted in blue and red color lines in **Figure 2**, and correspond to the node connections. The two red paths in **Figure 2** are the face routes which are passing through nodes closest to but never crossing the line connecting the source and the destination. In order to guarantee the loops-free paths, it is common to combine both of these basic strategies. For instance, a well-known GPSR protocol combines greedy forwarding with face routing.

It should be noted all the geographical routing protocols described in **Figure 2** assume unicast traffic, however, it is straightforward to extend these protocols for directional flooding. Similarly to multicast, geocast sends packets to a target group of nodes located in a given geographical area. There are also many geographical routing protocols for connected vehicles which exploit packet caching geographical maps of cities to predict the vehicle movements, for instance, GSR, GPCR, A-STAR, COIN, BREADCOMM, UMB and many others.

The location-aided routing (LAR) algorithm facilitates the geographical routing by partitioning the geographical area into two zones [8]. The expected zone narrows down the expected location of the destination. Such zone can be predicted from the past locations of the destination and information about the nodes mobility. The request zone defines the area where the search for a new route should be confined, for example, to flood the route request packet in order to significantly reduce the number of route-finding messages. If the packet is forward to a node outside the request zone, the packet is discarded. The LAR protocol can be combined with the greedy forwarding or directional flooding.

Geographical forwarding with expected zones is combined in the DREAM routing protocol. This protocol utilizes a position database where each entry contains a time-stamped information about the node current location, speed and direction in order to enable dead-reckoning predictions of location. The GRID routing protocol partitions the geographical area into a regular grid. At the local level, the packet routing is performed by some proactive routing algorithm whereas location-based routing is used to forward packets between the fields. The so-called home zone concept enforces all nodes within the home zone to keep information about the nodes belonging to that home zone, but which are temporarily away. In data centric networks, geographic location can be hashed to provide a unique key for data naming which also scales well in large networks.

4. Geographical RRM

Geographical RRM and multiple access schemes did not receive comparable attention as the geographical routing. Here, we partition the geographical region into non-overlapping areas referred to as virtual cells [13]. Unlike a partitioning into regular grid as for the GRID routing protocol discussed in the previous section, we define partitioning by the set of a priori chosen anchor points and partitioning is then form by the corresponding Voronoi regions. The virtual cells can be treated as the base station cells in cellular networks. It is then possible to pre-assign these cells with communication channels and other radio resources to facilitate distributed RRM and limit the exchange of control messages as well as to assume various channel reuse schemes. It is not necessary that the anchor points are static, or countably finite. The node clusters in MANET does not have to be fully contained within the virtual cells, although the virtual cells can be exploited to simply the clustering. In general, the utilization of virtual cells for RRM and multiple access is strongly dependent on the nodes mobility.

The RRM in wireless networks includes allocating communication channels, and setting the transmitting powers and data rates in order to use the limited radio resources as efficiently as possible. Unlike the cellular networks with centralized base station controllers, the RRM in MANETs is fully distributed, so the network nodes have to exchange enough information to coordinate multiple access, create network topology, manage interference, and determine routing. The scalability of MANETs is often achieved by a two-tier topology with clusters controlled by their respective cluster-heads. The nodes communicate with their cluster head who provide the centralized RRM within the cluster, however, the allocation of radio resources among the clusters remains distributed.

Virtualization of radio resources has recently emerged as a new paradigm to provide flexibility in efficiently sharing the network physical infrastructure. For instance, the network physical resources can be aggregated into a cloud, and then optimally partitioned to match the current demands of different users. It enables to define network function virtualization (NFV), virtual radio access networks (V-RAN), virtual operators and so on. Virtualization is expected to be the key design feature in the upcoming 5G networks. On the other hand, virtualization of the distributed radio resources in infrastructure-less networks is less straightforward, and it was rarely considered previously.

4.1 Virtual cells

The mobility model determines the optimum location of anchor points. Let the nodes in MANET follow the reference point group mobility (RPGM) model [14]. Such mobility consists of deterministic and random components. We assume that the random component represents a random waypoint (RWP) mobility, and for simplicity, the deterministic component representing a shared drift is the same for all nodes. For this mobility model, we can show that the optimum distribution of anchor points creates a hexagonal grid of equal-sized cells which is well known in the homogenous cellular networks. More precisely, the spatial mean of the RWP mobility is zero, i.e., such a node, on average, stays in one place. The optimum anchor point distribution is then given by the deterministic component of mobility, is also dependent on the initial placement of the network nodes. A large number of anchor points yield smaller virtual cells and more frequent handovers between them as the nodes move around. On the other hand, the virtual cells with large area may contain too many nodes, so the benefits of separating nodes into virtual cells diminish.

Assuming the deterministic component of the mobility is constant and the same for all nodes, and the random component of mobility for all nodes follows the same RWP model, the optimum anchor points lie on a rectangular grid with the dimensions $\sqrt{3}R/2$ and $3R/2$, respectively, where $R > 0$ is a scaling factor. The scaling factor is set to match the RWP model, i.e., the variance of the random mobility component in order to evenly distribute the nodes among the virtual cells. The anchor grid is rotated, so that it is aligned with the mean direction of the mobility. The corresponding anchor points are located in the 2D positions,

$$a_{m,n} = \left[Rm \bmod_2 (n-1), n\sqrt{3R/2}\right] \tag{1}$$

where m and n are integers. The virtual cells are defined by the Voronoi regions corresponding to the anchor points, and R represents the virtual cell radius. The list of anchor points is communicated to every network node. The nodes can determine in which virtual cell they are presently located by finding the closest anchor point. The virtual cells can be further sectored to aid the RRM.

4.2 Transmission channel allocation

The radio propagation model adopted greatly affects the performance of the network protocols. For our purposes to illustrate the concept of virtual cells, we assume that every node is equipped with an omnidirectional antenna. The transmitted signals are attenuated by independently and identically distributed fading coefficients drawn from the Rayleigh distribution. In addition, the signals are attenuated by the free-space path-loss modeled as,

$$\mathrm{PL}(d) = \mathrm{PL}_0 \times d^{-\mathrm{u}} \tag{2}$$

where d is the distance from the transmitter antenna, and $u > 1$ is the path-loss coefficient. The attenuation factor $PL_0 = \lambda_c/(4\pi)$ and λ_c denotes the carrier wavelength. The nodes are capable of full duplex transmissions, and they can transmit and receive at different frequency channels simultaneously.

As in the legacy cellular networks, the frequency channels can be reused in different virtual cells to increase the overall network capacity. The reuse distance for the hexagonal virtual cells defined by the anchor locations (1) is calculated as,

$$d_{\mathrm{reuse}} = R\sqrt{(3N_{cl})} \tag{3}$$

where $N_{cl} = (u^2 + v^2 + uv)$ is the number of cells in the cell cluster, and u and v are the number of cells which are crossed in order to arrive to the nearest co-channel cell within the hexagonal grid. Typical values of N_{cl} are 1, 3, 4, 7, and 9. In general, the larger the ratio d_{reuse}/R, the better the isolation between reused frequency channels, and the smaller the co-channel interference.

The cell coverage of the legacy cells and the proposed virtual cells are compared in **Figure 3**. In the former, the base station is at the cell center, so the cell radius R and the base station transmission range r are equal. In the latter, the transmission range r of the node at the virtual cell edge would have to be at least $r > 2R$ in order to cover the whole area of the virtual cell.

In order to assign the transmission channels, we assume there are F orthogonal frequency channels defined within the total bandwidth allocated to the network. Let F/N_{cl} be an integer, so the channels can be divided equally among the virtual cells in the same cell cluster. In order to provide the frequency diversity, the transmissions

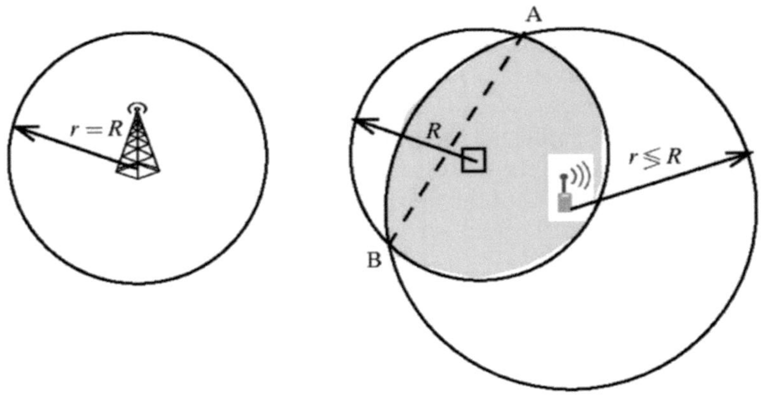

Figure 3.
The legacy cell coverage (left), and the virtual cell coverage (right). The square node represents the anchor point of the virtual cell.

adopt a frequency hopping patterns, so each node selects a different frequency channel for transmission at every time slot. The co-channel interference within the virtual cell is mitigated by defining a set of orthogonal frequency hopping patterns for the nodes in that cell. Note that these hopping patterns are only orthogonal as long as all transmissions are time-slot synchronized. Since the neighboring cells within the same cell cluster may not be time synchronized, we can reduce the resulting co-channel interference by requiring that every frequency channel is used within the cell cluster only once every $X > 0$ consecutive time slots. In particular, assuming $X = 3$, the channel allocation matrix F is resilient against the time-slot misalignment of the neighboring transmissions by up to one time slot. The following Matlab code generates the orthogonal channel allocation matrix F of $X \times N_{cl}$ frequency tones over T consecutive time slots for N_{cl} cells in the cell cluster. The resilience of the channel allocation matrix F will be shown numerically in the subsequent section.

The following Matlab code is used to generate the frequency hopping matrix F with the parameters X, N_{cl}, and T. The function randint (K) generates a random integer between 1 and K.

```
NA = zeros(X*Ncl,T); % auxiliary matrix
FF = zeros(Ncl,T); % channel matrix
for u = 1:Ncl
for t = 1:T
    i = find(NA(:,t)==0);
    j = randint(length(i));
FF(u,t) = i(j);
NA(i(j),t) = 1;
if t==T, t1 = 1; else t1 = t + 1; end
NA(i(j),t1) = 1;
if t==1, t1 = T; else t1 = t-1; end
NA(i(j),t1) = 1;
end
end
```

4.3 Geographical multiple access

Many general multiple access (MAC) link layer protocols were developed for MANETs such as Z-MAC, reservation MAC, distributed MAC, and spatial correlations based MAC. These protocols usually assume synchronized time-slots and carrier sensing to mitigate packet collisions. Here, we only consider a simple MAC scheme which can be supported by the virtual cells. We assume a two-tier network topology where the nodes are grouped in node clusters, so the packets are routed within and among the clusters. Each cluster elects a single cluster head node. The nodes connected to more than one cluster head serve as the gateway nodes between those clusters. The nodes can play other roles such as relaying the packets for other nodes as well as they can originate and consume traffic. We assume that each virtual cell is assigned a single frequency channel or a set of frequency hopping patterns. The node transmissions follow these rules:

1. The nodes in a given virtual cell can transmit only using the frequency channel or the frequency hopping pattern assigned to that cell. However, the nodes can listen to transmissions at multiple frequencies assigned to other neighboring cells.

2. The nodes within a given virtual cell use TDMA or mutually orthogonal frequency hopping patterns.

3. The nodes in different cells of the virtual cell cluster use FDMA or the assigned frequency hopping patterns.

In general, it is important to distinguish between the node clusters defined among the network nodes, and the cell clusters defined for the frequency channel reuse among the virtual cells. Consequently, the network clusters can be created independently of the nodes locations within the virtual cells. The virtual cells can contain nodes belonging to different node clusters, or there may be no cluster head within the virtual cell to time-synchronize the nodes and make their transmissions orthogonal. In order to overcome these issues and form the node clusters within the virtual cells, we assume the following assignment of the roles for nodes within the virtual cells:

1. The nodes located within the same virtual cell form a single cluster.

2. The node closest to the anchor point (i.e., the virtual cell center) becomes the cluster head.

3. The nodes at the edge between the virtual cells assume the roles of the gateway nodes for the other nodes in the cluster.

Choosing the cluster head close to the cell center leads to more efficient coverage of the cell, and smaller transmission distances from the other less centered nodes. The gateway nodes are selected to be close to the cell edge, and at the same time, they should be in different angular sectors. Since the packet relaying increases the number of transmissions in the cell, it should be limited. The role assignment for the nodes can be done by modifying the existing protocols used for creating the node clusters. Additional splitting of the virtual cells can be used to selectively poll the nodes in a predetermined order, for instance, the polling message requests the response from the nodes in a given cell sector. The node roles should be periodically updated as they move around. The node handover when leaving one cell and joining another cell can be performed by contacting the cluster head in the new cell and requesting the allocation of radio resources in that cell. The RRM performed by the cluster heads can be aided by exchanging location information of nodes in the same virtual cell.

4.4 Numerical examples

We assume that the anchor points are regularly distributed according to Eq. (1), and the cell radius $R = 500$ m. There are $N = 200$ nodes initially uniformly distributed in the observation rectangular area of 2500 × 2500 m. The deterministic component of the node movements is exactly horizontal whereas the random mobility component assumes the RWM model. As the nodes move around in the Eastern direction, their roles reestablished every 10 time slots. There is exactly one cluster head and up to three gateway nodes in each virtual cell. The gateways are the nodes furthest away from the cell center in each of the three sectors: 30 to 150°, 150 to 270°, and −90 to 30°, respectively. Hence, it is possible that, in some virtual cells, the cluster-head also acts as a gateway to transmit packets to the neighboring cells, otherwise, the cluster-head transmits packets to the nodes within the same cell. The remaining nodes in the virtual cells only retransmit packets to the other nodes within the same cell. The pairs of transmitting and receiving nodes in the virtual cells are chosen at random with a uniform probability. The pairs are selected independently from one time slot to another as well as independently among different cells. Thus, all transmissions within the same cell are orthogonal unlike the simultaneous transmissions in different cells. Furthermore, the transmissions assume frequency hopping where every cluster of the virtual cells is assigned a distinctive set of mutually orthogonal frequency hopping patterns. These patterns are generated by the algorithm presented in the previous section. Even though the transmissions in each cell at every time slot are orthogonal, the co-channel interference can still appear due to a lack of time-slot synchronization among the virtual cells, even within the same cell cluster. We assume the virtual cell clusters with $N_{cl} = 7$ and $N_{cl} = 3$ cells equal to the respective frequency reuse factors. The simulation snapshots for these two cases are shown in **Figure 4A** and **4B**, respectively. The arrows in these figures indicate randomly chosen transmissions. There are either one or two orthogonal transmissions per cell in each time slot.

We compare the following four transmission schemes. The first scheme, denoted as FR = 7/2 × 7, uses 2 × 7 = 14 distinct and orthogonal frequency channels with two of these channels allocated to every cell in the cluster of N_{cl} = 7 cells. Hence, there can be up to two simultaneous orthogonal transmissions in each cell in any given time slot. The frequency hopping pattern is created by randomly selecting two of the allocated frequency channels during each time slot. The second scheme, denoted as FR = 7/1 × 7, assumes seven orthogonal frequency hopping patterns over 3 × N_{cl} = 21 frequencies; one such pattern is allocated to each cell in the cell cluster. An example of these orthogonal patterns generated by the algorithm given in the previous section is presented in **Table 1**.

The third scheme, denoted as FR = 3/2 × 3, uses 2 × 3 = 6 distinct and orthogonal frequency channels the same way as the first scheme, but assuming only N_{cl} = 3 cells in the cell cluster. The fourth scheme, denoted as FR = 3/1 × 3, assumes three orthogonal frequency hopping patterns over 3 × N_{cl} = 9 frequencies which are generated and used the same way as in the second scheme.

The simulations in Matlab were performed to investigate the importance of time synchronization on the level of co-channel interference measured as the average signal-to-interference-plus-noise ratio (ASINR). The results for $T = 100$ time slots are shown in **Figure 5** assuming that the transmissions at the neighboring cells can be misaligned by up to one time-slot corresponding to $\Delta T = 100\%$. More specifically, given the value ΔT, the transmissions in $(N_{cl} - 1)$ neighboring cells are delayed by a fixed but randomly chosen time from the interval $(0, \Delta T)$. We observe that the

DOI: http://dx.doi.org/10.5772/intechopen.88110

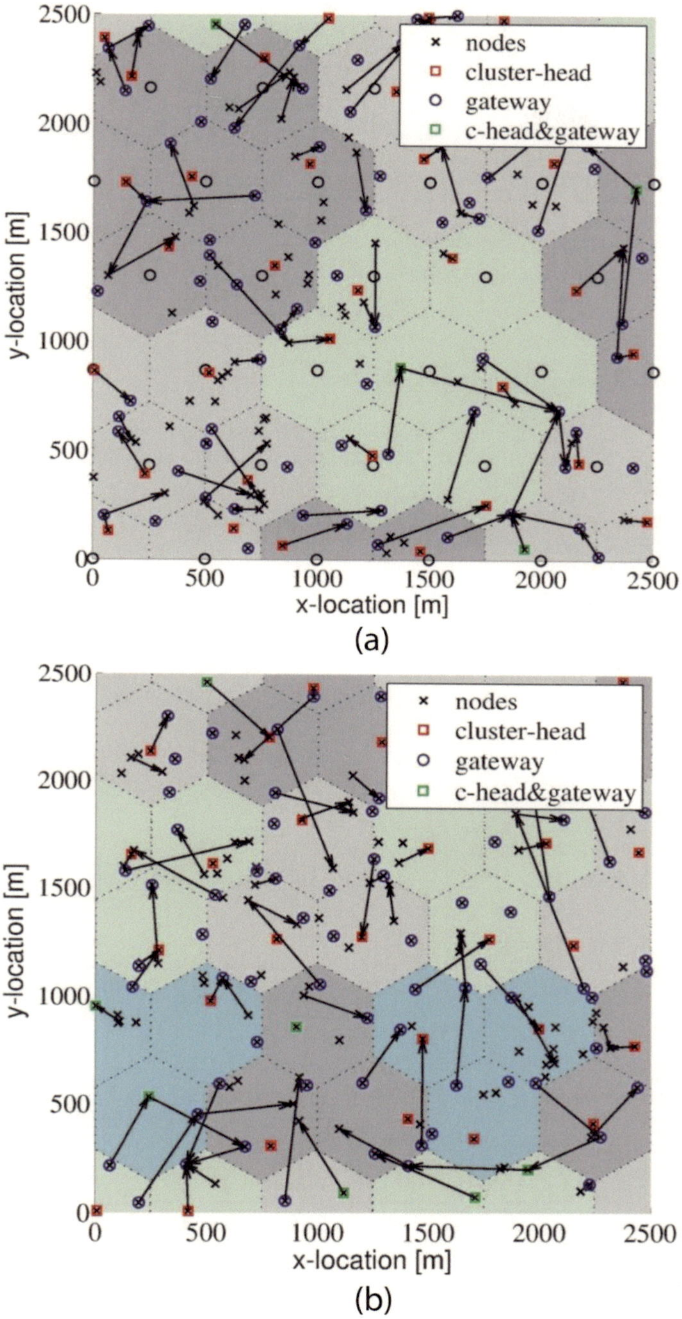

Figure 4.
(A) A snapshot of transmissions in the 7-cell cluster network. (B) A snapshot of transmissions in the 3-cell cluster network.

frequency hopping patterns generated by the presented algorithm are constrained such that the time delays by up to one time slot do not create any additional co-channel interference. On the other hand, the schemes 7/2 × 7 and 3/2 × 3 generate additional co-channel interference if the transmissions at subsequent time slots at the neighboring cells are occurring at the same frequency.

Cell	Frequency channels									
1	20	17	5	19	7	4	19	21	9	6
2	18	10	8	11	5	18	5	13	17	14
3	4	16	2	16	10	20	17	20	1	2
4	11	12	15	1	13	8	12	15	4	3
5	19	1	21	14	21	9	1	11	7	15
6	5	3	18	3	6	15	14	6	12	8
7	9	14	4	17	12	3	10	16	10	13

Table 1.
A sample orthogonal allocation of 21 frequency channels to a cluster of 7 cells over 10 time-slots.

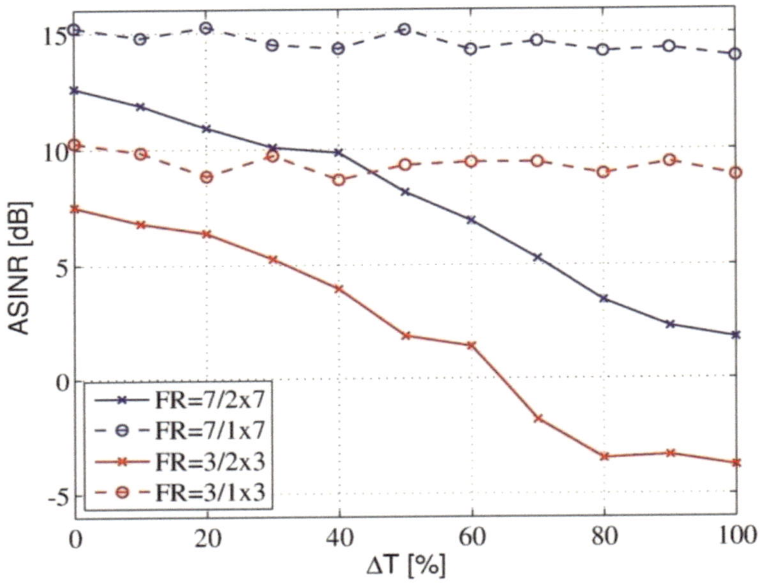

Figure 5.
The average SINR versus the timing difference ΔT for 4 channel allocation schemes.

4.5 Discussion

The localization methods including GNSSS consume additional energy, however, these methods are now routinely used in MANETs operating in the outdoor environments. The geographical partitioning of the area using a set of predetermined locations referred to as anchor points and the corresponding Voronoi regions can facilitate the frequency, or more generally, transmission channel planning. The reuse and assignment of communication channels in the cells is one of the main tasks of the base station controllers in the legacy cellular networks. Here, this task is accomplished without any supporting physical infrastructure, so the MANETs can take advantage of the virtually defined cells. The infrastructure-less virtual cells should be contrasted with the NFV and other virtualization strategies which are used to pool and partition the shared radio resources in radio access networks.

We illustrated the key concept of virtual cells, and how they can be used to facilitate distributed RRM and multiple access without any additional overhead. We made several simplifying assumptions, for instance, the deterministic component of the node mobility is aligned in one direction for all nodes, although the cell handovers and the reassignments of node roles were performed. The simulations were only concerned with the link layer protocols, but neither routing nor scheduling was considered, so traffic congestion in the network was not modeled. We investigated the transmission rules where the nodes can only transmit in the channels pre-assigned to the virtual cells whereas there was otherwise no restriction to which communication channels the nodes can listen to. We did not consider how the nodes can further exploit sharing their location information other than in determining their roles within the virtual cells. The interference due to asynchronous transmissions could be mitigated by employing spread-spectrum and multi-antenna techniques.

Much more sophisticated patterns of anchor points could be devised. The hexagonal regular cells are only optimum for very specific mobility model considered. Defining the optimum anchor points for general mobility models is an open research problem. Furthermore, the cluster heads in each virtual cell can adaptively request or advertise additional radio resources in collaboration with the neighboring cells. This could be triggered if the number of nodes in the cell goes above or below defined thresholds. The virtual cells can be adaptively adjusted by changing the number and positions of anchor points, or only some nodes may exploit virtual cells for RRM while other nodes operate under conventional RRM protocols. Such RRM strategies can better match the spatial node distribution over the area with virtual cells. Moreover, the node clusters may not be exactly contained within the virtual cells as considered in our simulations. In this case, a cluster head may be managing multiple virtual cells, or a virtual cell may be managed by multiple cluster heads. Another interesting problem is to investigate the co-existence of multiple MANETs in the area with defined virtual cells, the case of overlay multiple virtual cellular networks, and how to support virtual cells in the upcoming 5G networks. The geographical spectrum management can resolve many spectrum allocation problems.

5. Conclusion

The location information at nodes in wireless networks is becoming a commodity. It is likely that all transmissions in future wireless networks will be linked to exact geographical locations, and the wireless networks may be classified whether the use of location information at nodes is mandatory, optional or not used. The location information is readily available in outdoor environments using the GNSS service. The wireless localization techniques are more complicated to implement, and they consume bandwidth and energy. Hence, it is important to evaluate whether geographical protocols can outweigh these drawbacks. The key motivation for assuming geographical protocols is to improve the network efficiency while reducing the amount of overhead required for setting up and controlling network services, and to generally facilitate better mobility management.

We pointed out that geographical location is a trivial example of context which is naturally shared among the nodes in the network. It can be acquired internally within the network, for example, by means of collaborative localization techniques, or externally with the help of GNSS or terrestrial radio transmitters. There are works which exploit the GNSS timing signal to synchronize the nodes in the network. There may be other context types related to applications, devices or the network itself which can be exploited to improve the network protocols and services. This can be a fruitful area of research to explore in the 5G networks.

Geographical routing protocols were explained how they utilize geographical location information to improve the routing decisions and reduce the number of control messages. Both reactive and proactive routing protocols were discussed and advantages and disadvantages compared. The main challenge as in all other geographical protocols is how to efficiently acquire and share locations of all nodes in the network. This is equivalent problem to acquiring knowledge of all link costs in the network to facilitate optimum routing. Similarly as the link costs change in time, the nodes move and change their locations, so there must be some mechanism how to maintain up to date knowledge of node locations.

Unlike geographical routing protocols, geographical RRM and geographical multiple access did not receive comparable attention in the literature. The concept of virtual cells was proposed to facilitate RRM and multiple access in MANETs without requiring any additional overhead. The virtual cells are defined as Voronoi regions of spatially distributed anchor points. We assumed a two-tier network with clusters fully located in separate virtual cells, and designed a frequency hopping signaling scheme to maintain orthogonal transmissions within clusters of three and seven virtual cells, respectively. We also pointed out that, in all MANETs, the performance of all network protocols is directly affected by the radio propagation conditions, and the mobility of nodes. Finally, we outlined a number of open research problems to further develop the concept of virtual cells.

It should be noted that some topics mentioned in this chapter were treated rather superficially, for example, the mobility management and the localization mechanisms supported in the 4G/5G networks, since our main focus was on the geographical routing and geographical RRM. In addition, the most interesting and comprehensive research problem appears to be how to exploit the location information in MANETs where some mobile nodes are controlled by the 4G/5G network while the other nodes are autonomous and must perform the distributed routing as well as RRM using in-band or out-of-band signaling.

Acknowledgements

The parts of this book chapter were presented in a technical tutorial in the IEEE MILCOM 2018 conference in Los Angeles, CA on 29 October 2018.

Author details

Salman Al-Shehri[1], Pavel Loskot[1*] and Michael J. Hirsch[2]

1 Swansea University, Swansea, United Kingdom

2 ISEA TEK, Maitland, FL, USA

*Address all correspondence to: p.loskot@swan.ac.uk

IntechOpen

© 2020 The Author(s). Licensee IntechOpen. This chapter is distributed under the terms of the Creative Commons Attribution License (http://creativecommons.org/licenses/by/3.0), which permits unrestricted use, distribution, and reproduction in any medium, provided the original work is properly cited. (cc) BY

References

[1] Hayat S, Yanmaz E, Muzaffar R. Survey on unmanned aerial vehicle networks for civil applications: A communications viewpoint. IEEE Communications Surveys and Tutorials. 2016;**18**(4):2624-2661. DOI: 10.1109/COMST.2016.2560343

[2] Al-Shehri SM, Loskot P, Numanoğlu T, Mert M. Comparing tactical and commercial MANETs. In: Proceedings of the IEEE Military Communications Conference (MILCOM'17); 23-25 October 2017; Baltimore, MD, USA. 2017. pp. 1-6

[3] Roy RR. Handbook of Mobile Ad Hoc Networks for Mobility Models. US: Springer; 2011. p. 1103. DOI: 10.1007/978-1-4419-6050-4

[4] Borcea C. Mobile Computing [Internet]. 2017. Available from: https://web.njit.edu/~borcea/ [Accessed: 31 August 2018]

[5] Rahmati A, Zhong L. Context-for-wireless: Context-sensitive energy-efficient wireless data transfer. In: Proceedings of the 5th International Conference on Mobile Systems, Applications and Services (MobiSys'07); 11-13 June 2007; San Juan, Puerto Rico. 2007. pp. 165-178

[6] Wymeersch H, Lien J, Win MZ. Cooperative localization in wireless networks. Proceedings of the IEEE. 2009;**97**(2):427-450. DOI: 10.1109/JPROC.2008.2008853

[7] Kaur H, Singh H, Sharma A. Geographic routing protocol: A review. International Journal of Grid and Distributed Computing. 2016;**9**(2):245-254. DOI: 10.14257/ijgdc.2016.9.2.21

[8] Ko YB, Vaidya NH. Location-aided routing (LAR) in mobile ad hoc networks. Wireless Networks. 2000;**6**:307-321. DOI: 10.1023/A:1019106118419

[9] Cadger F, Curran K, Santos J, Moffett S. A survey of geographical routing in wireless ad-hoc networks. IEEE Communication Surveys and Tutorials. 2013;**15**(2):621-653. DOI: 10.1109/SURV.2012.062612.00109

[10] Grover J, Shikha, Sharma M. Location based protocols in wireless sensor network—A review. In: Proceedings of the International Conference on Computing Communications and Networking Technologies (ICCCNT'14); 11-13 July 2014; Hefei, China. 2014

[11] Kumar A, Shwe HY, Wong KJ, Chong PH. Location-based routing protocols for wireless sensor networks: A survey. Wireless Sensor Network. 2017;**9**:25-72. DOI: 10.4236/wsn.2017.91003

[12] Zhao L, Shen H. ALERT: An anonymous location-based efficient routing protocol in MANETs. In: Proceedings of the International Conference on Parallel Processing; 13-16 September 2011; Taipei, Taiwan. 2011

[13] Al-Shehri SM, Loskot P, Numanoğlu T, Mert M. Virtual cells for infrastructureless MANETs. In: Proceedings of the IEEE International Conference Communication, Information and Computing Technology (ICCICT'18); 2-3 February 2018; Mumbai, India. 2018

[14] Jayakumar G, Ganapathi G. Reference point group mobility and random waypoint models in performance evaluation of MANET routing protocols. Journal of Computer Systems, Networks, and Communications. 2008;**860364**:1-10. DOI: 10.1155/2008/860364

Section 3

Computing

Chapter 6

Architecture and Operation Algorithms of Mobile Core Network with Virtualization

Larysa Globa, Svitlana Sulima, Mariia Skulysh, Stanislav Dovgyi and Oleksandr Stryzhak

Abstract

The analysis of the current situation in the wireless communication market shows an increase in the workload, which leads to an increase in the need in additional resources. However, the uneven loading of the infrastructure nodes leads to their loss of use; so, there is a need in introducing technologies that both do not lead to downtime of equipment and ensure the quality of load service during the day. An overview of the NFV virtualization technology has shown that it is appropriate to build wireless networks, since it provides the necessary flexibility and scalability. The method for determining the location and capacity of reserved computer resources of virtual network functions in the data centers of the mobile communication operator, method for determining the size of computing resources constant configuration time interval, and distributed method of local reconfiguration of the virtual network computing resources in the case of a failure or overload are proposed. Thus, configuration, operation, and reconfiguration processes in mobile core network with virtualized functions are described.

Keywords: network function virtualization, evolved packet core, resource allocation, reconfiguration, mobile network

1. Introduction

In the mobile cellular network, the rapid development has been observed. Modern telecommunication systems are being constructed as complex networks that involve various types of devices united into a single complex, operating in conditions of large load flows and large number of connections [1]. They can offer higher data transfer rates, with the integration of more services and guarantee of high quality of experience. Nevertheless, this development also means that the amount of data that is transferred in the mobile network is increasing and the volume of signaling traffic is increasing, respectively. According to [2], it is expected that total mobile data traffic will have increased to 77 exabytes per month by 2022, almost seven times more compared to 2017. Mobile data traffic will grow at an average annual growth rate (CAGR) equal to 46% from 2017 to 2022.

According to Shimojo et al. [3], vehicles, houses, personal devices, robots, sensors, etc. will be connected wirelessly. It means that an automatic and intelligent control system will be achieved. An increase in the number of devices will affect the

 IntechOpen

IoT market, which is estimated to be $19 trillion [4], and is expected to reach 50 billion [5]. In addition, rich content services, such as real-time streaming movies that require high resolution and tele-surgery requiring small delay must be provided (**Figure 1**).

In addition, the average signaling requirement per subscriber is up to 42% higher in LTE compared to the standard of the past generation communication [6].

Furthermore, market competition requires faster deployment of services and elasticity of changing service criteria as well as the ability to cope with higher service requirements. Therefore, there is a need to manage the signaling traffic in order to provide the necessary quality of service to end users and the proper use of resources of the network operator.

In such circumstances, operators are forced to build up the network infrastructure to ensure the process of service of telecommunication services at a given level of quality. During the day, the load differs, and according to [7], up to 80% of the computing capacity of the base stations and up to half of the capacity of the core network are unused. This leads to a low usage of resources as well as a high level of energy consumption, which reduce the cost-effectiveness of the network for mobile operators.

The emergence of the concept of network functions virtualization opens up new opportunities for the world of telecommunication systems. At the same time, there is a need for new approaches, models, and methods for organizing service handling. The use of virtual servers to solve the tasks of the mobile core network can greatly simplify the process of organizing resources on the service server and ensure its scalability and fault tolerance.

The principle of network function virtualization (NFV) [8] is aimed at transforming network architectures by deploying network functions into software that can run on a standard hardware platform. According to the ETSI [9], the network function is a functional block within a network infrastructure that has defined external interfaces and a defined functional behavior. Network functions are components of the LTE evolved packet core (EPC) network, such as MME, HSS, PGW, and SGW, which for the NFV case will be deployed on the basis of data center system, with the use of leased computing resources (CPU core, memory, disk space, and network interface card), which can be allocated and reallocated in the process of operation depending on actual load requirements.

Thus, the features of NFV can be characterized as follows [10]:

1. Separation of software from hardware. Since the network element is no longer an aggregate of integrated hardware and software entities, the evolution of both is independent of each other. This allows having separate terms of development and maintenance of software and hardware.

2. Flexible deployment of network functions. The separation of software from hardware helps to reallocate and share infrastructure resources; thus, together

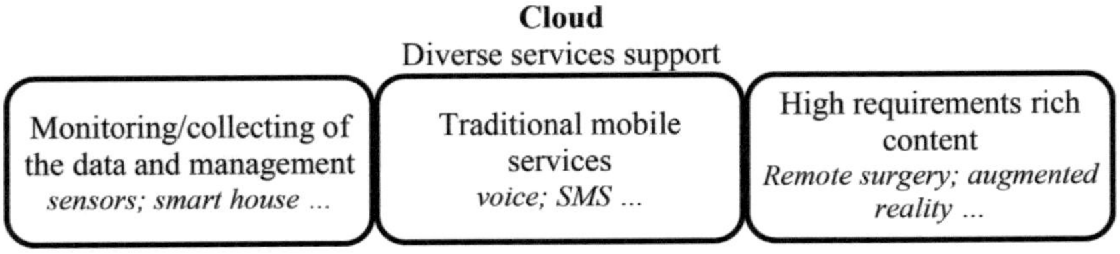

Figure 1.
Services in the era of future generations' networks.

hardware and software can perform various functions at different times. It helps network operators to deploy new network services faster on the same physical platform. Consequently, the components can be created in any NFV-compliant device on the network and their connections can be installed on a flexible basis.

3. Dynamic scaling. Dividing the functionality of the network function into created software components provides greater flexibility in scaling the actual performance of the virtual network function (VNF) more dynamically and with greater details, for example, according to the actual traffic for which the network operator should provide capacity.

At present, numbers of problems remain unresolved. You need to consider hybridity of the service environment, where flexible, well-scalable, virtual servicing entities located in rented cloud-based databases operate along with specialized hardware with limited features. Therefore, the task to organize the computing resources of service nodes and flows between them in a hybrid environment, which consists of hardware telecommunications and virtual computing entities, is important.

Unlike the existing static architecture of the LTE EPC network, a system (**Figure 2**) in which service flows are processed by hardware, and in the case of expected overload, the redistribution of flows happens and takes into account the expansion of the service network by adding virtual service facilities located in the leased clouds of the data centers is proposed (**Figure 3**). After organizing a hybrid service environment, there is a need to adapt the computing resources of the system in the process of operation to ensure a high-quality service, and also it is necessary to consider the features of the reconfiguration process and the costs associated with it. So far, there has not been any comprehensive solution to the task of controlling the computing resources of the hybrid telecommunication environment. The peculiarities of the load distribution of resources of network elements, hardware or virtual ones, have it been considered yet either.

Thus, the chapter proposes a structured approach to the management of resources of network functions through sequential control of the following stages:

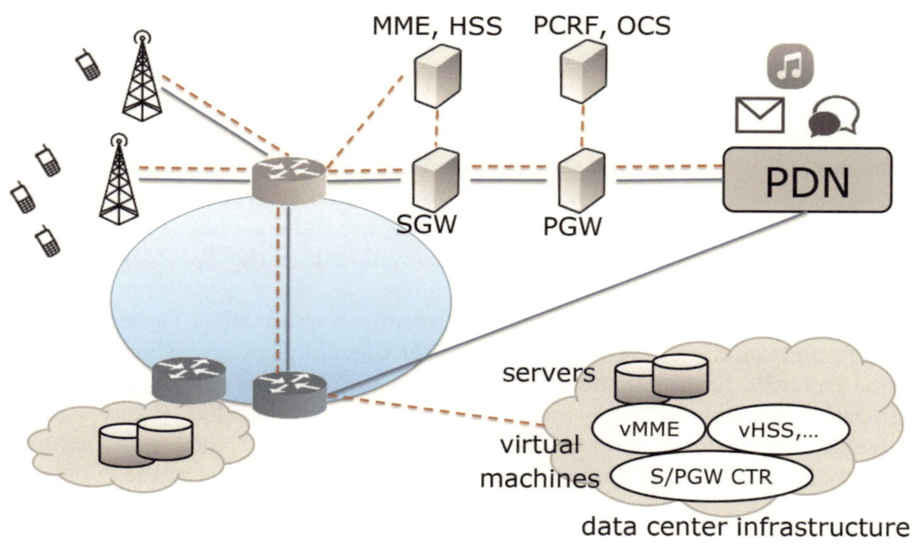

Figure 2.
LTE EPC network architecture with the use of NFV (variable dependent on load volumes).

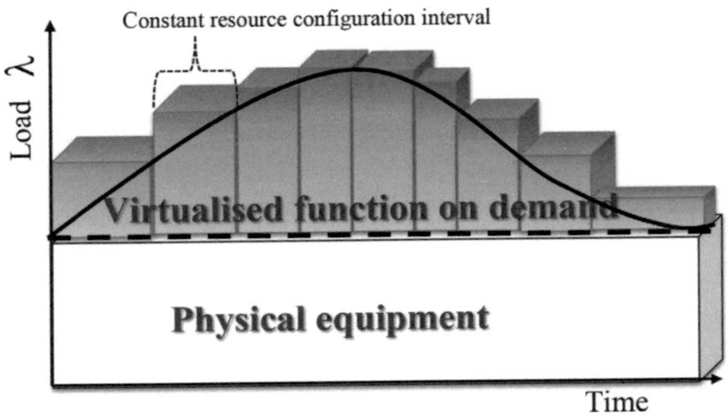

Figure 3.
Distribution of load in hybrid network.

monitoring, forecasting, controlling the sufficiency of resources, and controlling the given level of quality of telecommunication services.

Having analyzed the research and development processes of telecommunication networks of the next generation, we may argue that the existence of powerful data centers greatly expands the possibilities of organizing the process of providing services. One of the key aspects of network virtualization is the allocation of physical resources to virtual network functions. This involves mapping virtual networks on physical networks, as well as managing dedicated resources throughout the life cycle of a virtual network. The optimality and flexibility of resource allocation are key factors for successful network virtualization.

Most of the existing methods for solving the tasks of organizing hardware and virtual resources offer a static distribution of resources, in which, when computing and telecommunication environment is organized, the reallocation of resources does not occur throughout its life cycle. As the network traffic is not static, this may result in improper use of shared computing resources. It is important to organize monitoring of virtual nodes and provide the resources on the basis of their real needs.

2. Method for determining the location and required capacity of virtual reserved computing resources in case of an overload of the physical network

The method is based on the shared embedding concept [11] of the individual virtualized services of the core network on the physical network. We suppose that the virtual network functions of the mobile core network have the same functionality and interfaces as the network components of the 3GPP LTE EPC architecture.

The number of service chains must be determined in advance. The extreme case would be consideration of one service chain for the mobile phone/eNodeB. Since realistic scenarios for mobile networks are up to 10,000 eNodeBs, the resulting optimization model will be enormous and quite long computation time is required to solve it. Therefore, we accept reasonably large clusters of eNodeBs and assume that each of these eNodeB clusters refers to a single service chain of the core network.

Consider the situation when the provider of telecommunication services already has an existing topology of base stations. You need to determine a subset of the network nodes where the load aggregation blocks will be placed which will generate the requests to the same virtualized EPC service. After that, for each base station site, we assign a node of aggregation (traffic aggregation point – TAP).

Let x_i be a binary variable that is equal to 1 if we need to place TAP at point i and equals 0 in the other case. In addition, we define y_{ji} as a binary variable that is equal to 1 if the base station j sends the load to the i TAP and equals 0 in the other case. We need to define the values of x_i and y_{ji} in order to find the optimal value of the objective function.

Objective function (1) aims to minimize network latency. Objective function (2) represents the total cost of placing aggregation nodes and the cost of establishing channels between base stations and the respective TAPs. The objective function (3) aims to leave more free bandwidth on each physical channel. The residual bandwidth of all channels is maximized, since high-downloaded channels can lead to network overload; so, it is advisable to get a solution where more free channels are left.

These optimization goals can be useful for network operators to plan the best deployment strategy.

$$\min_{x_i,\, y_{ji}} \left(\sum_i \sum_j y_{ji} \cdot L_{ji} \right), \tag{1}$$

$$\min_{x_i,\, y_{ji}} \left(\sum_i x_i \cdot cost_i + \sum_i \sum_j y_{ji} \cdot costl_{ji} \right), \tag{2}$$

$$\max_{x_i,\, y_{ji}} \left(\sum_i \sum_j y_{ji} \cdot (c_{ji} - B_{ji}) \right), \tag{3}$$

where L_{ji} is the delay of the communication channel between the site j and TAP i;

$cost_i$ is the cost that consists of two parts: the fixed initial cost f_i, which is responsible for fixed investments such as space and installation of equipment, and the additional costs – $costN_i$ – per unit of processing power on the computing node, where d_i is the amount of computational resources of processing: $cost_i = f_i + costN_i \cdot d_i$;

$costl_{ji}$ is the cost of establishing a connection between the site j and TAP i, and it is determined as a linear combination of the initial fixed cost fl_{ji} and the variable part dependent on the bandwidth B_{ji}, which is necessary for the channel, and the cost of the unit of capacity $costLj$: $costl_{ji} = fl_{ji} + costL_i \cdot B_{ji}$;

c_{ji} – available bandwidth throughput.

It is possible to use the linear combination (4) of Eqs. (1)–(3) with weights a, b, and c, which can be applied not only to give importance the component but also in order to scale the values of these equations for the purpose of converting to comparable values and have meaningful summation:

$$\min_{x_i,\, y_{ji}} \left(\begin{array}{r} a \cdot \sum_i \sum_j y_{ji} \cdot L_{ji} + b \cdot \left(\sum_i x_i \cdot costN_i + \sum_i \sum_j y_{ji} \cdot costL_{ji} \right) - \\ -c \cdot \left(\sum_i \sum_j y_{ji} \cdot (c_{ji} - B_{ji}) \right) \end{array} \right) \tag{4}$$

Subject to:

$$\sum_i y_{ji} = 1\forall j, \tag{5}$$

$$y_{ji} \leq x_i \forall j \forall i, \tag{6}$$

$$\sum_i x_i \leq p, \tag{7}$$

$$\sum_j y_{ji} \cdot d_j \leq p_i \forall i, \tag{8}$$

$$\sum_i y_{ji} \cdot (c_{ji} - B_{ji}) \geq 0 \forall j, \tag{9}$$

$$\sum_i y_{ji} \cdot L_{ji} \leq T_j \forall j. \tag{10}$$

Restriction (5) ensures that each base station will be connected only one TAP. Restriction (6) ensures that a channel is created between the base station site j and the TAP i only if i was placed.

Restriction (7) ensures that the maximum TAP does not exceed budget p, while (8) is a capacity limit that ensures that the general requirements for processing of all base stations assigned to a specific TAP do not exceed the actual physical resources installed. Restriction (9) makes sure the sufficiency of channel resources for the establishment of channels, and (10) ensures admissibility of delay value, i.e., not exceeding the threshold T_j.

Below we describe the method for solving the problem of placement and the capacity of reserved computing resources of virtual network functions.

Physical network is given in the form of graph $SN = (N, NE)$, where N is a set of physical nodes and L is a set of channels. Each channel $l = (n_1, n_2) \in NE, n_1, n_2 \in N$ has a maximum capacity of $c(n_1, n_2)$ and each node $n \in N$ is associated with certain resources $c_n^{\,i}$, $i \in R$, where R is the set of resource types (CPU cores, memory, disk space, and network interface card). The set of all traffic aggregation points (TAPs), i.e., eNodeB clusters, is denoted as $K \subseteq N$. For each node $n \in N$, $suit_n^{\,k,j}$ is a binary parameter that indicates whether it is administratively possible to deploy a function $j \in V$ on the node n, where V is the set of types of network functions, k service, where $k \in K$.

A virtual mobile core network is represented by a set of services (one service per TAP) which are embedded in the physical network.

The requirements to the bandwidth between two functions, $j1$ and $j2$, $(j1,j2)$ E, referring to the TAP k K service are denoted as $d_k^{\,(j1,j2)}$. $d_k^{j,i}$ is the amount of computing resource type i allocated to the network function j in the service k. $s_{n,i}^{\;k,j}$ specifies the processing time for the type resource i of the virtual network function j for the service k with one resource unit on node n. The requirements to the admissible processing time of the network function j related to the service k are designated as P_k^j. T_k – the maximum delay for $k \in K$, $L(n_1, n_2)$ is the network latency for the channel $(n_1, n_2) \in NE$.

The goal of optimization is to find the location of the virtualized services of the core network (i.e., the allocation of network functions and the allocation of resources, as well as definition of the ways to transfer traffic between them), so as to minimize the cost of the occupied resources of channels and nodes in the physical network, while satisfying the load requirements $\lambda^{k,j}$. Let us formulate an objective function (Eq. (11)) in the form of a linear combination of two value expressions: the occupied capacity of computing node resources, where the value of resource unit i on node n is denoted by $costN(i,n)$, and the occupied bandwidth of the channels, where $costL(n_1,n_2)$ is the cost of the unit of bandwidth of the physical channel $(n_1, n_2) \in NE$.

The following Eqs. (11)–(20) represent the formulation of the optimization problem of mixed integer nonlinear programming. The Boolean variables $x_n^{k,j}$ indicate whether the network function j associated with the service k is located on the physical node n. For j = TAP, $x_n^{k,TAP}$ are not variables but input parameters that indicate where TAP k is, i.e.,

$$x_n^{k,TAP} = \begin{cases} 1 \text{ if } k = n, \\ 0 \text{ else} \end{cases}.$$

Similarly, Boolean variables $f_{(n_1,n_2)}^{k,(j_1,j_2)}$ indicate whether the physical channel $(n_1,n_2) \in NE$ is used for the path between j_1 and j_2 for service k.

$$\min_{x_n^{k,j}, f_{(n_1,n_2)}^{k,(j_1,j_2)}, d_t^{j,i}} \left(\sum_{k\in K}\sum_{j\in V}\sum_{n\in N}\sum_{i\in R} x_n^{k,j} \cdot d_k^{j,i} \cdot costN(i,n) + \sum_{(n_1,n_2)\in L} costL\left(n_1, n_2) \cdot \sum_{k\in K}\sum_{(j_1,j_2)\in E} f_{(n_1,n_2)}^{k,(j_1,j_2)} \cdot d_k^{(j_1,j_2)} \right) \right) \tag{11}$$

$$\text{Subject to } \sum_{n\in N} x_n^{k,j} = 1 \forall k\in K, j\in V \tag{12}$$

$$x_n^{k,j} \le suit_n^{k,j} \forall k\in K, j\in V, n\in N \tag{13}$$

$$\sum_{(w,n)\in NE}\sum_{k\in K}\sum_{(j_1,j_2)\in E} f_{(w,n)}^{k,(j_1,j_2)} \cdot d_k^{(j_1,j_2)} \le c_n^{bdw} \forall n\in N \tag{14}$$

$$\sum_{k\in K}\sum_{j\in V} x_n^{k,j} \cdot d_k^{j,i} \le c_n^{i} \forall n\in N, i\in \{R\backslash bdw\} \tag{15}$$

$$\sum_{k\in K}\sum_{(j_1,j_2)\in E} f_{(n_1,n_2)}^{k,(j_1,j_2)} \cdot d_k^{(j_1,j_2)} \le c(n_1,n_2) \forall (n_1,n_2)\in NE \tag{16}$$

$$\sum_{(n,w)\in NE} f_{(w,n)}^{k,(j_1,j_2)} - f_{(n,w)}^{k,(j_1,j_2)} = x_n^{k,j_1} - x_n^{k,j_2} \forall k\in K, n\in N, (j_1,j_2)\in E \tag{17}$$

$$x_n^{k,j}, f_{(n_1,n_2)}^{k,(j_1,j_2)} \in \{0,1\} \forall k\in K, j\in V, n\in N, (j_1,j_2)\in E, (n_1,n_2)\in NE \tag{18}$$

$$\sum_{(j_1,j_2)\in E}\sum_{(n_1,n_2)\in L} f_{(n_1,n_2)}^{k,(j_1,j_2)} \cdot L(n_1,n_2) \le T_k \forall k\in K \tag{19}$$

$$\sum_{n\in N} x_n^{k,j} \sum_{i\in R} \left(\frac{1}{\frac{d_k}{s_{n,i}^{k,j}} - \lambda^{k,j}} \right) \le P_k^{j} \forall t\in T, j\in V \tag{20}$$

Eq. (12) ensures that for each TAP/service, only one network function of each type is placed. Eq. (13) ensures that the allocation of resources is carried out on physical nodes, which have an administrative opportunity to locate the corresponding network functions. Eqs. (14)–(16) represent restriction for the available resources of physical nodes and channels. Eq. (17) represents a restriction for flow conservation of all paths in the physical network. Eq. (18) ensures that the variables in the task of locating network functions and displaying a path are Boolean.

In order to limit the delays on channels, the delay limit shown in Eq. (19) is also added. And to take into account the necessary performance of the virtual network function, the restrictions for the value of the processing time of the request determined in Eq. (20) are necessary.

It is supposed to solve the problem (11)–(20) in the offline mode at the initial stage. According to the solution, each network function reserves a certain number of resources of the virtual network function based on the assessment of its greatest resource requirements. The instantaneous needs of different network functions are dynamically satisfied by activating the necessary configuration of virtual machines during execution in such a way as to satisfy the guarantees provided for each network function.

3. Method for determining the size of the time interval of the constant configuration of computing resources

The decision when to provide resources depends on the dynamics of traffic loads. Telecommunication loads undergo long-term changes, such as hourly effects or seasonal effects, as well as short-term fluctuations such as unexpected crowds. While long-term fluctuations can be predicted in advance, observing changes in the past, short-term fluctuations are less predictable, and in some cases, unpredictable. The proposed method uses two different approaches for working in conditions of changes that are observed at different time scales. Proactive resource management is used to assess the load and corresponding management, as well as reactive resource management is used to correct long-term errors or to respond to unforeseen overload.

We propose to apply a mechanism which implies dynamic change in the duration of the constant configuration of the resources of the virtual network function, depending on the difference between the maximum load value at a certain base interval and the minimum one. Eq. (21) describes the principle:

$$\mathrm{Int(t)} = max\left(\begin{array}{l} Int_{base} \cdot \left(1 - K \cdot \dfrac{\max_{\tau\epsilon(t-I(t-1);t)}\lambda_{basepred}(\tau) - \min_{\tau\epsilon(t-I(t-1);t)}\lambda_{basepred}(\tau)}{\max\lambda_{basepred}}\right); \\ \hfill Int_{\min\ base} \end{array} \right), \tag{21}$$

Int is the interval during which the appropriate specified resources will be allocated; Int_{base} is the base value of the interval calculated according to the load discretization approach described below; *K* is the coefficient of the duration change of constant configuration determined by the network operator according to the experiment; $\lambda_{basepred}(t)$ is the average predicted arrival rate in the period *t*, and $Int_{minbase}$ is the minimum acceptable value of the base interval.

To do this, you need to define the base interval. The goal is to present a daily load pattern, sampling its requests into successive, non-overlapping time intervals with a single representative value in each interval. Load discretization: having a time series *X* in the interval $[v, \tau]$, time series *Y* on the same interval is the discretization of the load *X*, if $[v, \tau]$ can be divided into *m* consecutive non-overlapping time intervals, $\{[v, \tau_1],[\tau_1, \tau_2],...,[\tau_{m-1}, \tau]\}$, so that $X(j) = r_i$, for all *j* in *i*-th interval, $[\tau_{i-1}, \tau_i]$.

The solution for time series discretization (Eq. 22) is given as follows:

$$\sum_{i=1}^{m}\left[\sum_{\tau=\tau_{i-1}}^{\tau_i} u(r_i - X(\tau))\right] + f(m) \rightarrow \min. \tag{22}$$

Eq. (22) is an objective function which has to be minimized, where X is the time series and $f(m)$ is a function of the value of the number of changes or intervals, m. The purpose of Eq. (22) is to simultaneously minimize the load representation error and number of changes. Basic interval is calculated as $Int_{base} = \frac{\tau_m}{m}$. In order to determine the optimal value of the interval, we set different values of the number of intervals, calculate the value of the Eq. (22) and choose the best, i.e., the minimum one, having for each interval:

$$r_i = \max_{\tau\epsilon(\tau_{i-1},\tau_i)} X(\tau). \tag{23}$$

At the same time, it is proposed to continuously monitor the values of the request arrival rate and use the predicted values if the load does not exceed the threshold; otherwise, current trends are evaluated and resources are scaled on the basis of the new forecast.

Load forecasting for the next time interval is carried out by taking into account long-term statistics and adjusting it according to the model of exponential smoothing, where errors of more recent past periods have a greater importance factor:

$$\lambda_{pred}(t) = \lambda_{basepred}(t) + \alpha \sum_{j=t-h}^{t-1} (1-\alpha)^{t-1-j} \cdot \left(\lambda_{obs}(j) - \lambda_{basepred}(j)\right)^{+}, \tag{24}$$

α (smoothing constant) is the coefficient that characterizes the weight rate reduction and takes values from 0 to 1; the closer the value of this parameter is to 1, the better is the consideration of the influence of the last levels of the series during the forecast. The model parameters are set by the network administrator according to the experiment. $\lambda_{obs}(t)$ is the request arrival rate on interval t, h is the interval of previous observations, which is considered by the algorithm, and x^+ denotes $max(0,x)$.

4. Method of local reconfiguration of network computing resources in case of failure or overload

There might be situations when the resources available on the nodes will be insufficient or if the node fails. Potential failures can be physical nodes failures, failures of servers that have higher failure rates than telecommunication hardware or the infrastructure provider will perform node maintenance tasks and this will require the migration of nodes.

For this case, the methods of reconfiguration are used which seek to find the places for migration of network functions from the affected nodes, minimizing the cost of recovering the node after failure and maintaining a high level of physical performance of the network. The proposed improved recovery methods differ from existing ones by taking into account the cost of resources on the nodes and the final quality of service, as well as the case of node overload. In addition, in previous research, the problem of locating management nodes, which are coordinators of the movement of virtual network functions, remained unsolved.

MN represents a set of control nodes (hereinafter–managers), where managers *MN N* are responsible for the operation of the proposed recovery mechanism after the failure. Each control node is connected to one or more nodes in the physical network and performs the steps required to recover from the failure. Let us assume that managers can be located in nodes *N*. For a given number of managers *A*, there is a finite set of possible $\binom{|N|}{A}$ locations, so the task of placing managers is the task of multi-criteria combinatorial optimization. The purpose of the task is to determine the location of each manager at a given number of *A*, so that the general cost function $U_A(\{p_n{:}n\ N\})$ can be minimized, where p_n is a Boolean variable equal to 1 if the manager is placed at the point *n*. The task of optimization will be given as follows:

$$\begin{aligned} &\min_{\{p_n:n\in N\}} \ U_A \\ &\text{subject to} \sum_n p_n = A \end{aligned} \quad . \tag{25}$$

The main purpose of the optimal placement of managers is to minimize delays between nodes and managers in the network. However, considering only delays is not enough. The placement of managers should also take into account certain restrictions of stability. **Figure 4** shows different issues that need to be considered when evaluating the stability of the placement. Below we will briefly explain these issues and what is needed to be sustainable in relation to them. **Figure 4** shows normalized delays between nodes and arrival rate at nodes.

Let us presume that the nodes are assigned to their closest manager, using as the metric of delay, i.e., the shortest path $dl_{g1,g2}$ between the node *g1* and the manager *g2*. The number of nodes per manager may be unbalanced – the more nodes the manager has to control, the greater is the load per this manager. If the number of site requests to the manager in the network increases, additional delays probability due to the queues in the control system increases too. In order to be resilient to manager overload, the assignment of nodes to different managers should be balanced properly.

It is obvious that one manager is not enough to achieve network resilience. On the other hand, when multiple managers are hosted in the network, the logic of network management is distributed across multiple managers, and these managers must be synchronized to maintain a consistent global state. Depending on the frequency of synchronization between managers, the delay between individual managers plays an important role.

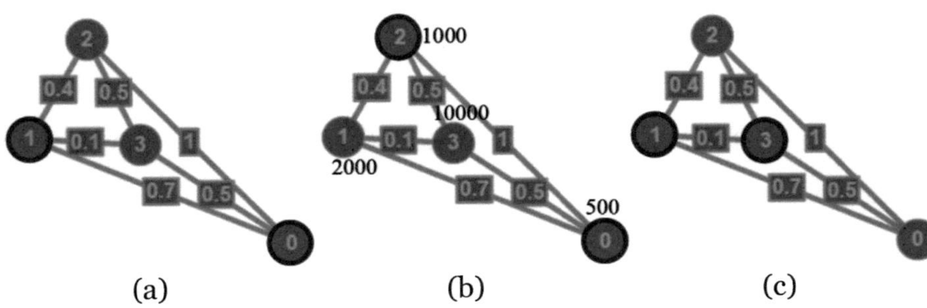

Figure 4.
Assignment according to different criteria: (a) minimal delay to the manager; (b) a minimum load imbalance of the managers; and (c) the minimum delay between managers.

Based on the *dl* matrix, which contains the distance of the shortest paths between all nodes, the maximum transmission latency between the node and the manager for a certain placement of managers can be calculated as follows:

$U_A^{latency}(p) = \max(ddc_n)$,

ddc_n is the maximum transmission delay from the network node to the manager at the point n;

ddc_n is calculated as follows:

$$ddc_n = \max_{g \in N} latency_g \cdot \pi_{g,n},$$

where $latency_g$ is the delay between manager and node g,
$latency_g = \min_{\{n:n \in N \cap p_n = 1\}} dl_{g,n}$;

$\pi_{g,n}$ is a Boolean variable equal to 1 if node g is served by a manager located at the point n.

We consider not the average, but the maximum delay value, since the average hides the values of the worst case which are important when resiliency needs to be improved.

Depending on the situation, it may be desirable to have an approximately equal load for all managers, so that no manager is overloaded, while others have little work. Next, we consider the balanced distribution of nodes between managers. As a formal metric, we introduce the balance of placement, or rather, the imbalance, $U_A^{imbalance}$, i.e., the deviation from the fully balanced distribution, as the difference between the load for the most downloaded manager and the least downloaded manager.

$U_A^{imbalance}$ is calculated as follows:

$U_A^{imbalance}(p) = max(ldc_n) - min(ldc_n)$, де $ldc_n > 0$,

ldc_n – manager load at n;

ldc_n is given as follows:

$$ldc_n = \sum_{g \in N} load_g \cdot \pi_{g,n},$$

where $load_g$ is the load factor for the node g.

As the last aspect of a resilient placement of managers, let us consider how the delay between managers can be taken into account when choosing managers. Formally, the delay between managers $U_A^{interlatency}$ is defined as the greatest delay between any two managers at a given placement:

$$U_A^{interlatency}(p) = \max_{\{n,g:n,g \in N \cap p_n = 1, p_g = 1\}} dl_{g,n}.$$

In general, placement with a delay between managers' considerations tends to place all managers closer to each other. This increases the maximum delay from nodes to managers.

Thus, the target optimization function is given by:

$U_A = wu^{latency} \times U_A^{latency}(p) + wu^{imbalance} \times U_A^{imbalance}(p) + +wu^{interlatency} \times U_A^{interlatency}(p)$,

where wu is the set of importance coefficients.

The recovery algorithm is based on prototype described in [12] but considers modified problem formulation and expands the solution on node overload case.

The physical network is given in the form of a graph $SN = (N,NE)$, where N is a set of physical nodes and NE is a set of channels. Each channel $(n_1,n_2) \in NE$, n_1, $n_2 \in N$ has a maximum throughput of $c(n_1,n_2)$ and a network delay $L(n_1,n_2)$, and each node n N is associated with certain resources $c_n^{\,i}$, $i \in R$, where R is the set of types of resources. The communication network is represented by a set of services (or virtual network requests) K that are embedded into the physical network. The virtual network request k, $k \in K$, can be given as a graph $G_k = (V_k, E_k)$, where V_k is the set of virtual nodes containing h_k elements and denoted as $V_k = (v_{k,1}, v_{k,2}, \ldots, v_{k,hk})$, where $v_{k,j}$ indicates the *j-th* network function in the service chain of k. E_k is the set of virtual channels $e_k(v_{k,j}, v_{k,g}) \in E_k$. The channel throughput requirements between the two functions, $j1$ and $j2$, referring to the $k \in K$ service are marked as $d_k^{(j1,j2)}$, $d_k^{j,i}$ is the number of resource type i allocated to the network function j in the *k-th* service. The Boolean variables $x_n^{k,j}$ indicate whether the network function j associated with $k \in K$ is located on the physical node n, and the variables $f_{(n1,n2)}^{k,(j1,j2)}$ determine whether the physical channel $(n1, n2)$ is used in the path between $j1$ and $j2$ for request k. L_k is the maximum delay for request k. $costN(i,n)$ is the cost of the occupied resource unit on the physical node n, and $costL(n_1,n_2)$ is the cost per unit occupied bandwidth on the

$x_n^{k,j} \leftarrow 0$
$S_1 \leftarrow \{m : \exists e_k(j,m)\}$
for all $\{m \in S_1\}$ *do*
  $f^{k,(j,m)} \leftarrow 0$
  $w_m \leftarrow M_N(v_{k,m})$
end for
$S_2 \leftarrow \cup_{m \in S_1} w_m$
Manager sends SPT request to all physical nodes in S_2
for all $w \in S_2$ *do*
  Perform SPT algorithm
  $S_{3,w} \leftarrow \{q : length(q,w) \leq l\}$
end for
$S_4 \leftarrow \varnothing$
for all $q \in \cup_{w \in S_2} S_{3,w}$ *do*
  for all $\{m \in S_1\}$ *do*
    if $\exists e_k(j,m)$ *then*
      $f_{(q,w_m)}^{k,(j,m)} \leftarrow 1$
    end if
  end for
  if $(\sum_{(b_1,b_2) \in E_k} \sum_{(a_1,a_2) \in NE} f_{(a_1,a_2)}^{k,(b_1,b_2)} \cdot L(a_1,a_2) \leq L_k$ && $d_k^{j,i} \leq c_q^{\,i}$ $\forall i \in R$
  $d_k^{(j,m)} \leq c(q,w_m)$ $\forall m \in S_1)$ *then*
    $CostNL_q \leftarrow \sum_{i \in R} d_i^{k,j} \cdot costN(i,q) + \sum_{w \in S_2} d_k^{(j,m)} \cdot costL(q,w_m)$
    $S_4 \leftarrow S_4 \cup q$
  end if
  for all $\{m \in S_1\}$ *do*
    $f^{k,(j,m)} \leftarrow 0$
  end for
end for
if $S_4 = \varnothing$ *then*
  Perform Reconfiguration
else
  Select min $CostNL_q$, $q \in S_4$
  $q^* = \text{argmin } CostNL$
end if
$x_{q^*}^{k,j} \leftarrow 1$
for all $\{m \in S_1\}$ *do*
  if $\exists e_k(j,m)$ *then*
    $f_{(q^*,w_m)}^{k,(j,m)} \leftarrow 1$
  end if
end for

Figure 5.
Algorithm of recovering the node with a failure.

physical channel $(n_1,n_2)\in NE$. $suit_n^{k,j}$ means that the j function k can be placed on node n.

The process of moving the nodes of the virtual network hosted on the failed node, $v_{k,j}^{fail}$, starts when the system sends a recovery request to the corresponding host manager. The recovery process for each failed virtual node proceeds as follows: the manager sends the recovery request to all nodes of the physical network, which hosts the virtual nodes adjacent to the affected virtual nodes. Each of these nodes builds the shortest path tree (SPT) to all nodes of the physical network at a distance of not more than l (the threshold is set by the service provider) from the node, where the SPT root is the node. The manager uses these paths to select the node with the optimal distance to all nodes in the physical network, where the nodes of the virtual network are located adjacent to the failed node. This node eventually becomes the best candidate for hosting the affected virtual host. In addition, the capacity of the end nodes of the paths with the SPT should be at least the capacity of the virtual node located on the failed node. We select a node with a minimum cost of the path to all root nodes in the SPT trees and the minimum processing cost. **Figure 5** contains a description of the pseudo-code of the recovery algorithm (**Figure 6**) after failure and is applied for all $\{v_{k,j}: x_n^{k,j} = 1 \;\&\; n = failed\}$.

There is also a probability of the node failure due to overload. To perform a recovery in an overloaded network, the reconfiguration procedure is performed to migrate the virtual nodes hosted on the overloaded physical node.

The recovery process begins with sorting all the virtual nodes located on the overloaded physical node. The criterion (CRT in **Figure 7**) used to sort these nodes in a virtual network is the capacity of the virtual nodes. Then, the recovery

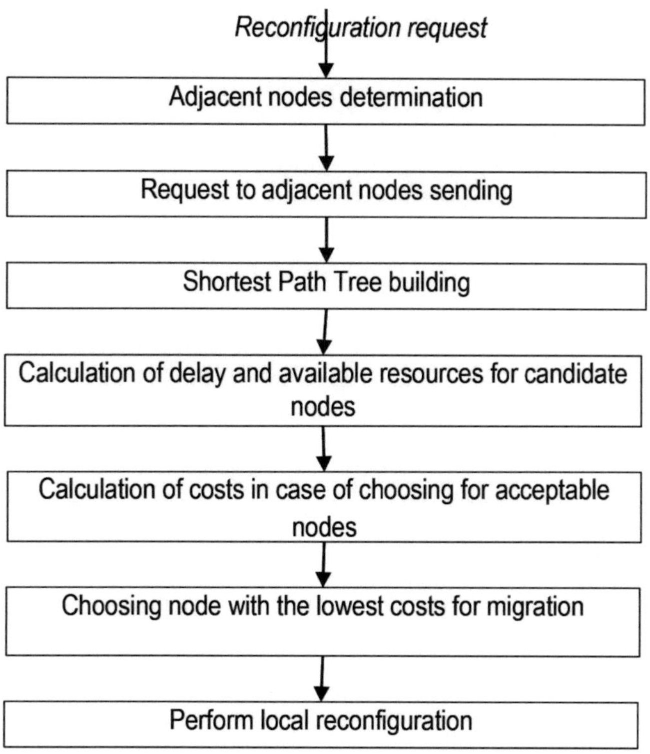

Figure 6.
Recovering the node with a failure.

n=overloaded
S_1←Sort virtual nodes on node n in ascending order based on CRT
Select from S_1 first virtual node v_{kj} with resource capacity not less then overloaded capacity
$d_k^{j,i} \geq \Delta c_n^{\ i}\ \forall i \in R$
Perform Node Recovery algorithm

Figure 7.
Recovering the node with overload.

procedure is performed on the first sorted virtual network node, which has a capacity equal to the overloaded, to migrate to the new node of the physical network.

When the load or resources change, some virtual network functions (VNFs) may have to be moved. There is a probability that finding a new node candidate for a node of a virtual network hosted on a failed site will not be possible. In this case, the reconfiguration procedure is performed to migrate one or more virtual nodes. Let us consider the problem of migration as an optimization problem, which is aimed at minimizing the general migration costs with the limits of permissible delay and computational resources.

The goal of optimization is to find the location of virtual network functions (i.e., the location of network functions and resource allocation as well as channels to transfer traffic between them), so as to minimize the cost of the occupied resources of channels and nodes in the physical network, while satisfying the requirements of traffic. Let us give the objective function (26) in the form of a linear combination (with weighted coefficients a, b, c, and e) of the cost expressions.

Let us determine the binary variable $x_n^{k,j} \in \{0,1\}$ to indicate that VNF j is associated with the service chain k placed on the node n after migration. The indicator $x_n^{k,j} = 0$ means that VNF j is not placed on node n after migration; otherwise, j is placed on node n after migration.

Then, we enter the binary variable $y_n^{k,j}$ to display the network status before migration. It is similar to variable $x_n^{k,j}$, $y_n^{k,j} = 0$ means that VNF is not located on node n before migration; otherwise, j is located on node n before migration.

Thus, we can use the $I^{k,j}$ indicator to indicate whether the VNFj by k service was moved in the current migration process.

$$I^{k,j} = \sum_{n \in N} x_n^{\ k,j} \cdot y_n^{\ k,j}$$

$I^{k,j} = 0$ indicates that the VNF has been moved in the current migration process, and $I^{k,j} = 1$ indicates that the VNF has not been moved.

x_n denotes whether the n physical server works or not after migration.

$$x_n = \begin{cases} 1 \text{ (server n launched)} \\ 0 \text{ (otherwise)} \end{cases}$$

y_n indicates whether the n physical server works or not before migrating.

$$y_n = \begin{cases} 1 \text{ (server n launched)} \\ 0 \text{ (otherwise)} \end{cases}$$

In order to consider the resources that are consumed while migrating, we introduce the following equations:

B_n indicates the required b_n costs to launch the *n-th* server:

$$B_n = b_n x_n (x_n - y_n);$$

$L_i^{k,j}(n \to n')$ denotes the use of resource i for the migration of VNF j from the service chain k from the server n to the server n':

$$L_i^{k,j}(n \to n') = l_i\left(d^{k,j}\right) + l_i'\left(d^{k,j}\right),$$

where $l_i(x)$ is the function of using resource i for migration from the server and $l'_i(x)$ is the use of resource i for migration to the server.

The objective function will be calculated as follows:

$$MCost = a \cdot \sum_{n \in N} (B_n + x_n \cdot cost(n)) + b \cdot \sum_{n \in N} \sum_{k \in K} \sum_{j \in V} \sum_{i \in R} x_n{}^{k,j} \cdot d_i{}^{k,j} \cdot costN(i,n) + c \cdot \sum_{(n_1,n_2) \in NE} costL(n_1,n_2) \cdot \sum_{k \in K} \sum_{(j_1,j_2) \in E} f_{(n_1,n_2)}{}^{k,(j_1,j_2)} \cdot d_k{}^{(j_1,j_2)} + e \cdot \sum_{n \in N} \sum_{n' \in N} L_i^{k,j}(n \to n') x_{n'} (x_{n'} - y_n) \tag{26}$$

Taking everything into account, we formulate the problem as follows.

Objective function:

Min MCost.

With constraints:

$$\sum_{n \in N} x_n{}^{k,j} = 1 \forall k \in K, j \in V, \tag{27}$$

$$x_n{}^{k,j} \le suit_n{}^{k,j} \forall k \in K, j \in V, n \in N, \tag{28}$$

$$\sum_{k \in K} \sum_{j \in V} x_n{}^{k,j} \cdot d_i{}^{k,j} + y_n{}^{k,j} \cdot (1 - I^{k,j}) \cdot l_i\left(d^{k,j}\right) + x_n{}^{k,j} \cdot (1 - I^{k,j}) \cdot l'_i\left(d^{k,j}\right) \le c_n{}^i \forall n \in N, i \in R, \tag{29}$$

$$\sum_{t \in K} \sum_{(j_1,j_2) \in E} f_{(n_1,n_2)}{}^{k,(j_1,j_2)} \cdot d_k{}^{(j_1,j_2)} \le c(n_1,n_2) \forall (n_1,n_2) \in NE, \tag{30}$$

$$\sum_{(n,w) \in L} f_{(w,n)}{}^{k,(j_1,j_2)} - f_{(n,w)}{}^{k,(j_1,j_2)} = x_n{}^{k,j_1} - x_n{}^{k,j_2} \quad \forall k \in K, n \in N, (j_1,j_2) \in E, \tag{31}$$

$$x_n{}^{k,j}, f_{(n_1,n_2)}{}^{k,(j_1,j_2)} \in \{0,1\} \forall k \in K, j \in V, n \in N, (j_1,j_2) \in E, (n_1,n_2) \in NE, \tag{32}$$

$$\sum_{(j_1,j_2) \in E} \sum_{(n_1,n_2) \in NE} f_{(n_1,n_2)}{}^{k,(j_1,j_2)} \cdot L(n_1,n_2) \le L_k \forall k \in K, \tag{33}$$

$$\sum_{n \in N} x_n{}^{k,j} \sum_{i \in R} \left(\frac{1}{\frac{d_k}{s_{n,i}{}^{k,j}} - \lambda^{k,j}} \right) \le P_k{}^j \forall k \in K, j \in V \tag{34}$$

Hence, the objective function (26) is a linear combination of four equations which aims to minimize: the cost of starting and using a server, using server resources, communication channels, and resources for migration. Eq. (27) ensures the one-time allocation of network functions, and Eq. (28) is the administrative possibility of placement on the node. Eqs. (29) and (30) represent a limit for the resources of physical nodes and channels, i.e., they ensure that the amount of resources involved in a node does not exceed the amount of available resources. Eq. (31) represents a flow conservation limit, i.e., the input stream at the node is equal to the output stream. Eq. (32) ensures that the variables in the problem are Boolean. Eqs. (33) and (34) represent a limit for the time of transmission by telecommunication channels and time of processing by service nodes, respectively, and ensure compliance with the specified time requirements for the service.

5. Operating scheme of the resource management system

Thus, before operation starting, it is necessary to have statistics on the requests arrival rate for the network function and the probability characteristics of the request servicing. According to the allocation method, the binding of each network function of the traditional network to the data center and the amount of resources that should be reserved for the corresponding virtualized network function is determined. Next, it is necessary to divide the lifecycle of the network function into intervals during which its configuration will remain unchanged and a certain amount of resources will be activated in accordance with the method of determining the size of the resources constant configuration time interval, while taking into account the expected load. When a mobile network operates, a physical node may not be able to continue to handle an incoming load due to lack of resources or due to its failure, and in this case, a distributed local reconfiguration of resources that re-distributes virtual nodes is triggered.

The general resource management system is shown in **Figure 8**.

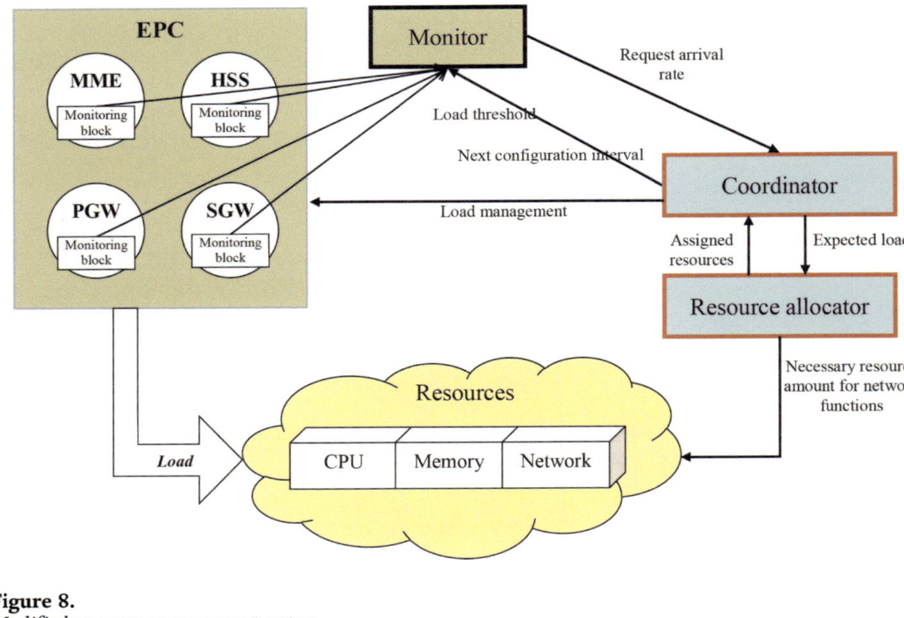

Figure 8.
Modified resource management system.

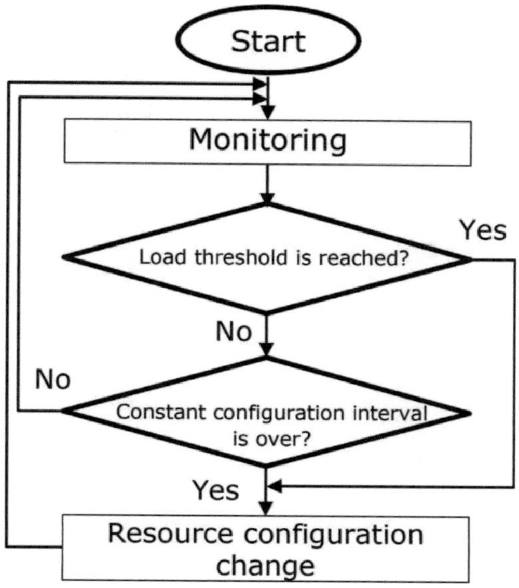

Figure 9.
The method of system resource management.

The monitoring system tracks traffic and counts the number of requests. The monitoring system sets the threshold for the number of requests and sends a message to the coordinator if detecting an overload. When the coordinator accepts an overload message from the monitoring system, the resource allocation unit calculates the required amount of resources to process the applications properly and dynamically distributes the estimated volume. Then, the coordinator redirects the requests and the overload is eliminated.

The coordinator is launched periodically. To predict the base load, one can take the average value of historic daily load. The coordinator sends an incoming load to a data center, which maintains excessive workload, and also exchanges data with a resource allocation unit to provide information about the predicted input load.

The resource distribution module is responsible for distributing the appropriate amount of resources needed to handle the load with the specified quality indicators. During the direct operation of the system, this module is started when the actual load exceeds the base predicted value of the load in order to provide additional resources for excessive load. Since the resource distribution module and coordinator do not start when the actual load is lower than the predicted one, the resource reconfiguration procedure creates minimal additional costs associated with this process.

The general operating scheme of the resource management system is shown in **Figure 9**.

6. Analysis

Quantitative and qualitative analysis (**Figure 10**) of the proposed methods showed a reduction of the cost associated with reserved resources up to 15%, which contributes to increasing the efficiency of load processing, saving computing resources.

The examples of representation of time series values, i.e., loads that illustrate the accuracy of representation, depending on the selected interval of the constant

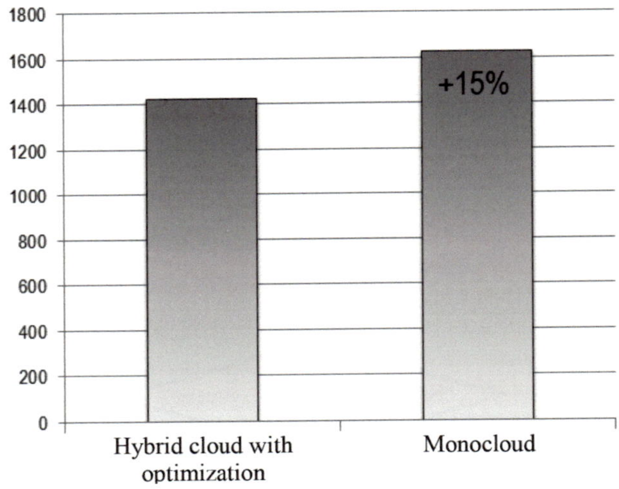

Figure 10.
Consumption of system resources by using the method of allocating virtual network functions and without it.

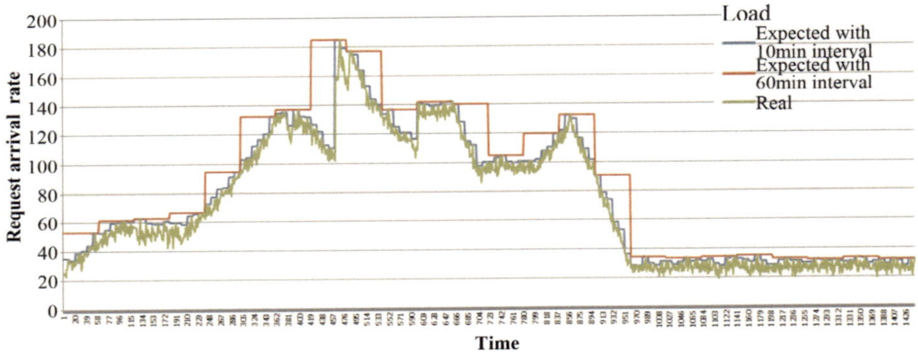

Figure 11.
Representation of load values depending on different values of the constant configuration interval.

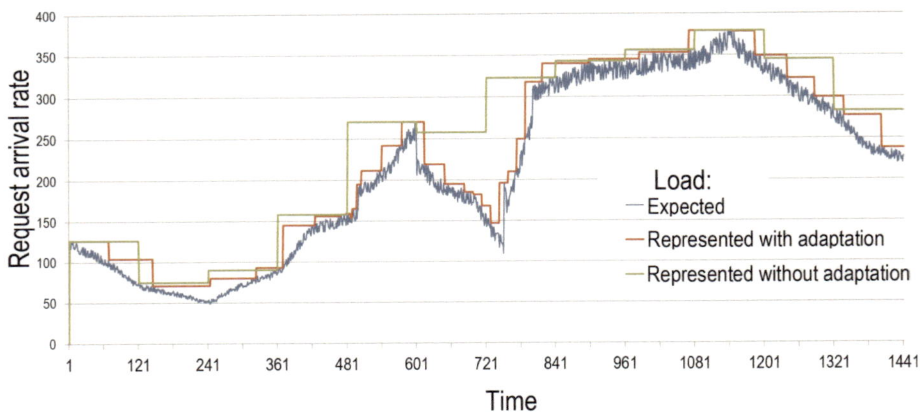

Figure 12.
Results of simulating the system with dynamic change in the value of constant configuration time interval and the system without it.

configuration, are presented in **Figure 11**, where the representation error for the case of intervals in 10 minutes is 7%, and for the case of 60 minutes—19%.

The results of simulation of the method of determining the size of the resources constant configuration time interval (**Figure 12**) showed that the difference between representational value and actual one can be 9%. If you do not apply a dynamic adjustment system to the value of the constant configuration interval, then the deviation will be 18%, i.e., 9% more, and the resources will be spent more.

In order to assess the proposed approach, the average amount of free resources per day was determined as the difference between fixed allocations, i.e., when 100% of resources were always allocated during the day, and dynamically allocated resources by using NFV. According to the results of simulation, the volume of resources allocated dynamically on average is 42% less than in the case of using the traditional distribution approach. **Figure 13** depicts the result of the dynamic distribution of resources in the virtualized EPC of the mobile network in a graphical form. A gray line illustrates the fixed allocations for the worst case scenario. The black curve shows the amount of resources distributed dynamically according to the proposed method.

According to the simulation results (**Figure 14**), the proposed local reconfiguration method showed up to 27% lower costs compared to a strategy aimed at minimizing delay, the delay being within the permissible limits but by 20% greater.

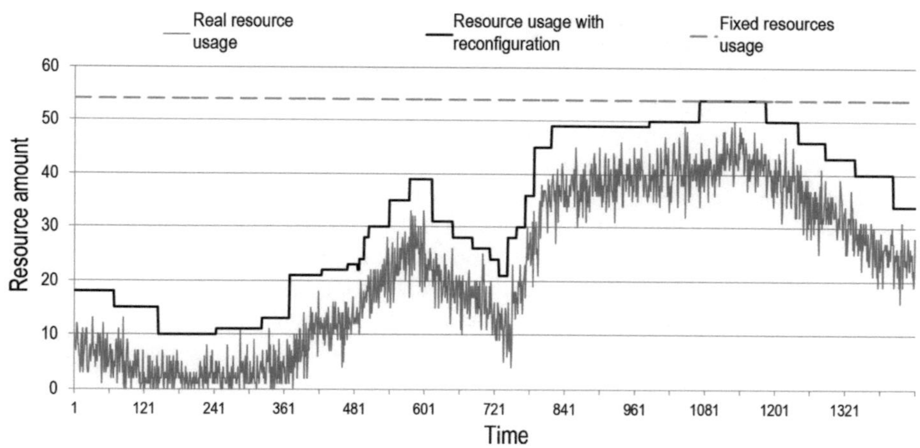

Figure 13.
Results of simulating the system with variable configuration of resources and the system with fixed resources.

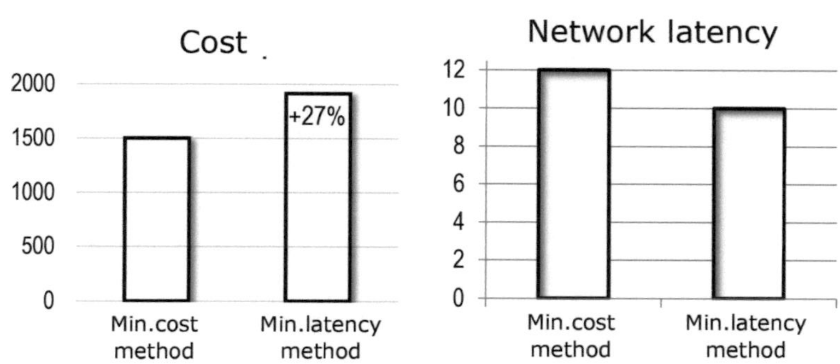

Figure 14.
Results of simulating the system with variable configuration of resources and the system with fixed resources.

7. Conclusions

The main result of the study has become the development of the method for reconfiguring resources of the core network by means of virtualization technology. As a result of the research, the following basic scientific results have been obtained.

The analysis of the current situation in the wireless communication market shows an increase in the workload, which leads to an increase in the need in additional resources. However, the uneven loading of the infrastructure nodes leads to their loss of use; so, there is a need in introducing technologies that both do not lead to downtime of equipment and ensure the quality of load service during the day.

An overview of the NFV virtualization technology has shown that it is appropriate to build wireless networks, since it provides the necessary flexibility and scalability.

We have developed the method for determining the location and capacity of reserved computer resources of virtual network functions in the data centers of the mobile communication operator, which guarantees the quality of providing telecommunication services with the minimum necessary resources by determining their sufficient configuration in a heterogeneous environment of available resources. This allows reducing costs by 13% compared to the randomly selected monocloud and by 47% compared with the traditional approach to deploying the network.

In addition, we have developed the method for determining the size of computing resources constant configuration time interval, which involves its changing and the consideration of both the cost of reconfiguration and the use of resources, as well as provides a flexible use of resources in the virtualized environment, which reduces the percentage of free resources by 42% compared to the dedicated equipment and by 9% compared to existing analogs and reducing the workload on the network.

Furthermore, we have improved the distributed method of local reconfiguration of the virtual network computing resources in the case of a failure or overload, which uses decentralized management and considers migration costs, that redistributes virtual network functions in normal and emergency modes while providing rational resource usage and reducing costs on average by 21%.

Author details

Larysa Globa[1], Svitlana Sulima[1*], Mariia Skulysh[1], Stanislav Dovgyi[2] and Oleksandr Stryzhak[2]

1 National Technical University of Ukraine "Igor Sikorsky Kyiv Polytechnic Institute", Kyiv, Ukraine

2 National Academy of Sciences in Ukraine, Kyiv, Ukraine

*Address all correspondence to: itssulima@gmail.com

IntechOpen

© 2019 The Author(s). Licensee IntechOpen. This chapter is distributed under the terms of the Creative Commons Attribution License (http://creativecommons.org/licenses/by/3.0), which permits unrestricted use, distribution, and reproduction in any medium, provided the original work is properly cited.

References

[1] Globa L, Kurdecha V, Ishchenko I, Zakharchuk A, Kunieva N. The intellectual IoT-system for monitoring the base station quality of service. In: Proceedings of the IEEE International Black Sea Conference on Communications and Networking (BlackSeaCom); 4-7 June 2018; Batumi, Georgia: IEEE. 2018. pp. 1-5

[2] Cisco Visual Networking Index: Global Mobile Data Traffic Forecast Update, 2017–2022 White Paper [Internet]. 2019. Available from: https://www.cisco.com/c/en/us/solutions/collateral/service-provider/visual-networking-index-vni/white-paper-c11-738429.html [Accessed: 31 March 2019]

[3] Shimojo T, Takano Y, Khan A, Kaptchouang S, Tamura M, Iwashina S. Future mobile core network for efficient service operation. In: Proceedings of the IEEE Conference on Network Softwarization (NetSoft); 13-17 April 2015; London, UK: IEEE. 2015. pp. 1-6

[4] Cisco CEO at CES 2014: Internet of Things is a $19 Trillion Opportunity [Internet]. 2014. Available from: https://www.washingtonpost.com/business/on-it/cisco-ceo-at-ces-2014-internet-of-things-is-a-19-trillion-opportunity/2014/01/08/8d456fba-789b-11e3-8963-b4b654bcc9b2_story.html [Accessed: 31 March 2019]

[5] Emmerson B. M2M: The internet of 50 billion devices. WinWin Magazine. January 2010. pp. 19-22

[6] Signaling is Growing 50% Faster than Data Traffic [Internet]. 2012. Available from: http://docplayer.net/6278117-Signaling-is-growing-50-faster-than-data-traffic.html [Accessed: 31 March 2019]

[7] NSN to Push Cloud Computing to Telco Gear Market [Internet]. 2011. Available from: https://www.reuters.com/article/us-nokiasiemens-gear/nsn-to-push-cloud-computing-to-telco-gear-market-idUSTRE78I6LK20110919 [Accessed: 31 March 2019]

[8] Network Functions Virtualisation [Internet]. 2012. Available from: https://portal.etsi.org/NFV/NFV_White_Paper.pdf [Accessed: 31 March 2019]

[9] Network Functions Virtualisation (NFV); Terminology for Main Concepts in NFV [Internet]. 2014. Available from: https://www.etsi.org/deliver/etsi_gs/NFV/001_099/003/01.02.01_60/gs_nfv003v010201p.pdf [Accessed: 31 March 2019]

[10] Mijumbi R, Serrat J, Gorricho J, Bouten N, De Turck F, Boutaba R. Network function virtualization: State-of-the-art and research challenges. IEEE Communication Surveys and Tutorials. 2015;**18**:236-262. DOI: 10.1109/COMST.2015.2477041

[11] Baumgartner A, Reddy V, Bauschert T. Mobile core network virtualization: A model for combined virtual core network function placement and topology optimization. In: Proceedings of the IEEE Conference on Network Softwarization (NetSoft); 13-17 April 2015; London, UK: IEEE. 2015. pp. 1-9

[12] Abid H, Samaan N. A novel scheme for node failure recovery in virtualized networks. In: Proceedings of the IEEE International Symposium on Integrated Network Management (IM 2013); 27-31 May 2013; Ghent. Belgium: IEEE; 2013. pp. 1154-1160

Chapter 7

Mobile Distributed User Interfaces

Erika Hernández-Rubio, Amilcar Meneses-Viveros and Sonia G. Mendoza-Chapa

Abstract

The success of a mobile application is due to the usability that the graphical user interface provides. A feature of mobile devices is the limited space for the interaction and the deployment of the graphical user interface. For this reason, user interfaces can have different interaction modalities. However, to work with information that can be complex to display, the use of modalities may not solve this problem. A possible alternative to provide more workspace to the users is through a distributed user interface (DUI). A mobile DUI allows the mobile applications to use two or more devices to execute the user interface. These devices can be Smart TVs or wearable such as smart watches. In this work the concepts of mobile DUI design are discussed, some use cases are presented and it is shown that its development in mobile devices is feasible.

Keywords: distributed user interfaces, mobile application, mobile devices, plasticity

1. Introduction

In the last years the use of mobile devices has increased. The success of mobile devices is due, among other factors, to its moderate cost, the variety of applications that allow being connected to the Internet, and the ease of use for many of its applications [1, 2]. The usability of the applications of the mobile devices is the main characteristic for the acceptance of the users [2]. This implies that the applications are intuitive and easy to use. To achieve this, researchers and developers have proposed design guides, patterns and templates to achieve applications with good features and easy to use. In addition, due to the diversity of sensors that mobile devices have, they can have different interaction modes and gestures that are used to control applications [3, 4].

Despite all the innovative technological elements for a pleasant user experience presented by mobile devices, for the user they have the restriction of the size of the screen, which reduces their area of work. The range of displays for smartphones is between 4 and 7 inches [5]. The sizes of the tablets are between 7 and 18 inches. And the range for smartwatches is between 1.2 and 2 inches. While there are mobile devices with screens larger than 13 inches, most of these devices are below 10 inches [5].

To get the most out of the work area offered by most mobile devices it is needed to take advantage of the work areas of the different mobile devices that the user has, depending on the context in which the user is. To achieve this, it is possible to design applications for mobile devices that can work with DUIs. That is, an application takes advantage other mobile devices carried by the user or other devices such as Smart TV

IntechOpen

in the area where it is located. Another problem in DUIs design is the quality in order to guarantee the usability and functionality of applications that use DUIs [3].

User interfaces have a time component that allows establishing whether the adaptability of their elements will be done dynamically or statically. Dynamic adaptability refers to the changes that the graphical interface makes when the application detects a change of context. Static adaptability is established when the user chooses how the graphical interface will adapt before doing a task or when starting a session. Therefore, several researches have developed concepts such as DUI and plasticity of user interfaces.

In this work we present concepts of DUIs, plasticity, and mobile computing to establish the specific restrictions for DUIs of mobile applications, and to discuss how the plasticity concepts of user interfaces complement the handling of these restrictions to establish the concept of mobile DUI. We present the design methods that have appeared in the literature and emphasize that both are complementary to a mobile DUI design.

2. Mobile distributed user interfaces

A mobile DUI is a DUI that takes advantage of mobile devices, communication networks and context of use to distribute user interfaces to take advantage of the display restrictions of mobile devices. It should be clear the concepts of DUIs and the characteristics of mobile applications have a complete notion of mobile DUI.

The use of DUIs is very common in multimedia applications such as music players, video players, image galleries, video games, books or interactive learning materials, but there are still few applications that use it for purposes other than entertainment. DUIs can be used in educational contexts [6]and for assistance applications for disabled persons [6, 7]. Also DUIs are required to interaction with smart spaces [8–11].

One approach to understand the use of DUIs is in [2], where the authors discuss the evolution of trends of computing since main frames to ubiquitous computing (UC). With the arrival of UC, users interact with more computing devices that contain input and output elements. In this section we start with a discussion of the DUIs, then a discussion of the characteristics of the mobile applications, and finally a discussion is presented to define the concept of mobile DUI.

2.1 Distributed user interfaces

A user interface (UI) is the set of elements that allow the user to interact with computers. These elements can be categorized as input, output and data control. This definition involves all kinds of technology and interaction mechanisms.

Vanderdonckt [12] propose a transversal model to distribute the user interface across users, platforms and environments. In this model, the authors consider the triplet $C = (U, P, E)$, where U is the user model, P is the platform model, and E is the environment model. Vanderdonckt considers these three elements the dimensions for UI distribution (**Figure 1**).

With this model it is possible to determine what elements of the UI would be distributed, to know the interaction modality that will be used when the elements are ported

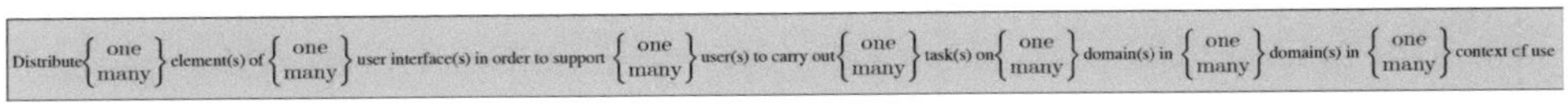

Figure 1.
Transversal model of DUI.

to target platform, to know the tasks that will be performed in a lapse of time, to know the domains involved in the distribution, and to know the platforms that participate in that distribution configuration.

A distributed user interface is a set of UIs that can be implemented in more than one device, or software platform. Some implementations consider the use of two or more devices simultaneously [7, 13]. By authors Penaver, Melchior and Gallud in several papers from 2011 to 2013 [14–17] we know that any single user interface can be cataloged as a distributed user interface if it has some characteristics like portability, fragmentation (also known as decomposition), simultaneity, and continuity. Being the first two characteristics the most important to satisfied the transformation of a user interface to a distributed one [14, 17, 18].

Portability. Means that a user interface can be completely or partially transferred in order to achieve a better user interaction.

Fragmentation. Any user interface can be fragmented, only if its different fragments can be run independently without losing functionality.

Simultaneity. If a user interface can run in different, software or hardware, platforms and also can be managed at the same time, it means that the UI is a simultaneous system.

Continuity. This characteristic is reachable when a system element can be moved to another module. This element is also a part of our distributed user interface, but always preserving its state.

From 2011 to 2013 several authors make some definitions [14–18] to formulate the DUI abstract model that allows developers to arrive at an implementation model. In this model, the elements of interaction (input, output and control), functionality, target, user interface, portability, decomposability, sub-user interfaces, platform, distributed user interfaces, simultaneity, requirements function and concurrency restriction stand out.

2.1.1 Definitions

In [7–9], the authors present a mathematical formalization of the DUI to obtain the properties of portability, fragmentation, simulation and continuity. This formalization is based on the next definitions:

Interaction element: An interaction element $e \notin E$ is an element that allows a user u to make an interaction in a platform p. An element can be an input element, an output element or a control element.

Functionality: Two interactions elements $e_1, e_2 \notin E$ have the same functionality if a user performs the same action using them.

Target: A subset of elements $E_0 \subset E$ have the same target if $\forall e_i \in E_0$ a user gets an action of a target task using the element functionality.

User interface: An User Interface $S_0(I)$ is a set of elements that have the same target. From [14, 15], a User Interface is defined by a set of interaction elements that can perform a task in a specific context.

Platform: An interaction element $e \notin E$ exists in a platform $p \notin P$ if e is supported, implemented or executed on p. Furthermore, a user interface I is supported in p if $\forall e \notin I$, then a user u can perform an interaction using e on p.

User subinterface: Let I be a graphical interface that allows a user u to reach a target T on a platform $p \notin P$. If the target reached is a subtarget of T, then the set of elements that is associated with the subtarget form a graphic subinterface.

Distributed user interface: A distributed user interface $DI \in DUI$ is defined as a user interface which has been fragmented and ported. DUI is a set of interaction elements that come from a set of subinterfaces.

State of user interface: The state $S(I)$ of a user interface I is defined as the set of values or modes associated to the interaction elements and target in a user interface after the user has reached a target associated with I. Every user interface I_i has an initial state $S_0(I)$ that changes when an element of I is used to make an interaction of u with p.

State of distributed user interface: The state of a distributed user interface $S(DI) = (S(I_1), \ldots, S(I_n))$ is a n-tupla where each element is the state of the user interface I_i that makes up the DUI.

With these definitions it is possible to have a formal description of the characteristics for distributed user interfaces such as portability, fragmentation, simultaneity and continuity. These characteristics are very important for working with distributed user interfaces. Villanueva et al. [3] propose the use of these characteristics as metrics to determine the quality of DUIs.

2.2 Mobile applications

A mobile application is an application that runs following the mobile computing paradigm. In this paradigm, the application's view layer runs on a mobile device, and the business and storage layers may or may not be on or off the device. In addition, the device must have the ability to be always connected (anywhere at any time) taking advantage of the different infrastructures of communication networks and also must consider the mobility of the user [19]. Mobileness means that the use of an application is always under an environment with constant changes, so the application must be able to adapt to changes in the context to remain functional and usable to the end user.

A mobile application is a computer mobile software designed to perform a task or to provide a user experience. The mobile software development presents some special requirements [20]:

Interaction with other applications: most of the mobile devices have many applications from different sources. New applications should be able to interact with the installed applications.

Sensor handling: the applications must be able to use the device sensors in order to improve user experience.

Families of hardware and software platforms: most embedded devices execute code that is custom-built for the properties of that device, but mobile devices may have to support applications that were written for all of the varied devices supporting the operating system, and also for different versions of the operating system.

User interfaces: they must be usable. The design of a user interface must consider the device's constraint like display size, battery life and processor capacity, and it must take advantage of the device's capabilities.

Power consumption: many aspects of applications affect the use of the device's power and thus the battery life of the device. Mobile applications may make extensive use of battery.

2.3 Plasticity of user interfaces

Techniques for reconfiguring the components of an application must be used. In addition to these concepts, it should be considered that a user interface in a mobile environment would be affected by changes in the environment, so the term plasticity turns out to be relevant. The plasticity of user interfaces is their ability to adapt to the context of use and to preserve their usability [21, 22]. This concept is useful to handle the adaptation of the elements in a DUI. Due to the mobility and ubiquity inherent in this type of systems, the changes of context are natural in this type of systems [23].

The context deals with the evolution, the structuring and the exchange of information spaces [24], which are designed to fulfill a particular purpose. In plastic user interfaces, the purpose is to support the process of adapting the user interface to preserve usability, i.e., plasticity techniques must handle the context of use. A change of context could be defined as the modification of any element of the contextual information space.

Vanderdonckt et al. [23] define seven dimensions to manage plasticity: adaptation means, UI component granularity, state recovery granularity, UI deployment, context of use, technological space coverage, and plastic meta-UI.

2.4 Mobile DUI

In the literature there are several models to design a DUI and give quality. We can notice that these models complement each other. Vanderdonckt's transversal model uses three dimensions, Penalver's, Melchoir's and Gallud's model uses four dimensions, and Vanderdonckt's plastic model considers seven dimensions. The work that has been done with DUIs and their formalization establish that the elements of a UI can be distributed, and the relationship that exists between them. Those works on plasticity of user interfaces establish how adaptability can be made, focusing on the conditions of context.

To handle the context for a mobile DUI, it must be considered that the information spaces are the elements of the UI (elements, sub UI, etc.), and the characteristics of the devices where the DUI will be displayed. The plasticity of the UI must handle the context of use. A change of context is the set of devices where elements of the user interface can be displayed, and in this way it is observed that elements of the information space are modified.

In general, it is possible to distinguish two methods for DUI design. One of these is presented by [12] and the other is presented by [14–17]. In [12] the way to distribute the GUI elements between users, platforms and environments is emphasized. In [14–16] the design is considered through a conceptual model based on portability, decomposition, simultaneity, and continuity of the DUI. The formalization of the DUI helps to know what elements are going to be distributed. Plasticity helps to establish when to make the distribution of elements and also raises the problem of how to do it.

We can say that a mobile DUI is a set of DUIs that mainly uses mobile devices sensitive to context, which can be supported by ubiquitous computing. The use of mobile devices allows the use of the different interaction modalities that they include. However, DUIs must consider handling the restrictions of mobile devices, mainly the size associated with the user interface, the high dependence on network connectivity and battery management.

There are several mobile platforms such as Android, iOS or Windows Mobile, among others, that provide tools and frameworks for developing applications. Furthermore, kits and frameworks have been developed to create mobile applications that allow us to share displays among mobile devices of the same family, i.e., their frameworks allow us to build some DUIs. Every platform uses a different strategy because the development paradigm is different for each platform. Another option is to develop rich clients that run on a Web browser.

However, despite the advantages offered by mobile application development platforms, it is necessary to use middleware and frameworks that help to efficiently manage DUIs. Another problem remains the design of the UIs that will be distributed to each platform involved in the DUI.

Work has been done to have models for the design, development and deployment of DUIs in execution time [8, 14, 25–28]. These works consider software engineering techniques as well as aspects of implemention. These last considerations can be reinforced with the works on plasticity that have been developed [29, 30].

3. Examples

With the elements described in the model of Section 4, we present three examples where the design of the DUI is available in three combinations of computing devices: Tablet-Smart TV, smartphone-smartwhatch, and Tablet-Tablet.

3.1 Tablet-SmartTV

In this example, an application is presented to perform three neuropsychology tests of Luria: Poppelreuter I, Poppelreuter II and Raven [7]. This application is called LuTest, whose architecture allows the user to manage a DUI whose platforms are an iPad tablet and an Apple TV, as shown in **Figure 2**. The main task of this application is to apply to a user the neuropsychological tests of Poppelreuter I, II and Raven. This example presents two ways to show the elements of the output: one is to duplicate the UI and the other to divide the UI. The final users of these applications are older adults, so the design of the UI is aimed at this population. Applications can only run on the Tablet or they can use the Tablet-Smart-TV combination in order to increase the work area.

DUIs have a static adaptability. The user determines the tablet orientation and the mode of work: alone or with a Smart TV before starting the test. The designs of the UI, in all cases of adaptability, are oriented to work with older adults. Because a dynamic adaptability could generate confusion in final users, we decide make a static adaptability.

3.1.1 DUI properties for Poppelreuter I and II

In Poppelreuter I test, the user must indicate the figures presented to them with visual noise. Poppelreuter 1 test begins by showing the user the contour image of an object, later more images containing the original object will be shown, but now the outline is combined with lines that may confuse the patient. The user must indicate the outline of the original object, ignoring the additional lines. The test consists in displaying different images, with different objects and different types of visual noise, as shown in **Figure 2**. For Poppelreuter II test, the visual noise is generated by overlapping the contours of several forms, **Figures 2** and **3**. The user must indicate the contour for each of them.

Portability: In Poppelreuter I and II there is partial portability. In the tablet the UI is maintained and in the Smart TV the output elements are deployed. These elements display the contours of the figures. In addition, the user can notice the progress of the test and the type of color they have selected to identify each figure separately.

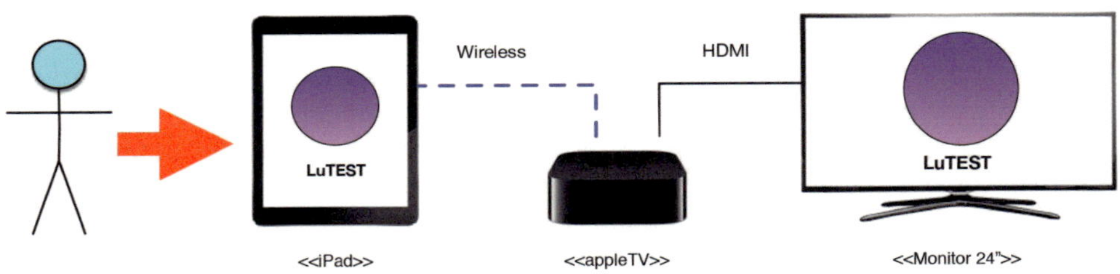

Figure 2.
Architecture for LuTEST. This application makes a DUI using a Tablet iPad and an Apple TV.

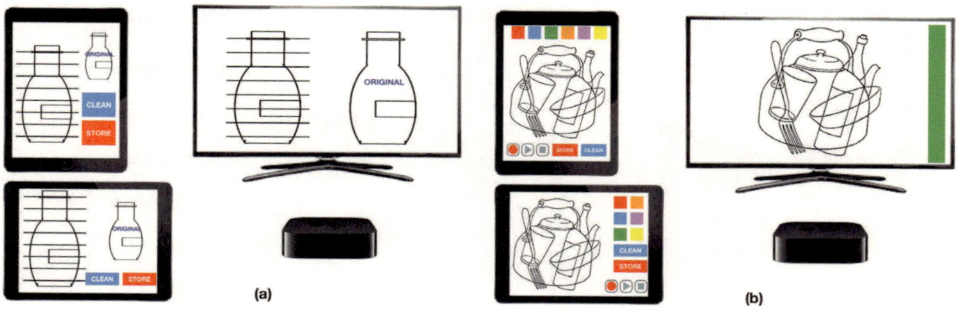

Figure 3.
(a) DUI design for Popperreuter 1 test using a tablet in landscape or portrait orientation with a Smart TV and (b) DUI design for Popperreuter II test using a tablet in landscape or portrait orientation with a Smart TV.

Fragmentation: The initial UI is divided into two UIs. The UI 1 contains the elements that the contour figures display and allows interaction through the touch screen. The UI2 contains the elements such as buttons and color palettes that the user can choose to perform the test.

Simultaneity: When the user works with the input elements found in the UIs of the Tablet, the status changes are reflected on the Smart TV in real time.

Continuity: This property is not essential in this example, because the UI enters an initial state when distributing the user interface. Due to the requirements of the application, the elements of the user interfaces do not move during the application of the neuropsychological tests.

3.1.2 DUI properties for Raven test

The Raven test is used to evaluate visual and cognitive abilities. It works as follows: the patient observes a certain visual structure, which is incomplete. The patient can choose between six or eight possible options, but only one is correct. In some cases, the patient is asked to differentiate their answers from the others, and for that the patient must grasp the principle under which each option was constructed. The complete Raven test is composed of three series, each with 12 different test matrices whose difficulty progresses step by step. The advantage of using Raven to assess cognitive abilities is that a grammatical knowledge or a complex mathematical ability is not required (**Figure 4**).

Portability: The UI is partially transferred from the Tablet to the Smart TV. The input elements are maintained in the Tablet and the output element is sent to the Smart TV.

Fragmentation: The UI is fragmente in two sub UI. The UI1 has all input elements. The UI2 has the output element.

Simultaneity: When the user works with the input elements found in the UIs of the Tablet, the status changes are reflected on the Smart TV in real time.

Continuity: This property is not essential in this example, because the UI enters an initial state when distributing the user interface. Due to the requirements of the application, the elements of the user interfaces do not move during the application of the neuropsychological tests.

3.2 Smartphone-smartwatch

In this example, a DUI is presented, which allows the communication of smartwatch and smartphone type mobile devices. The objective of this DUI is to show the best walking route that a tourist should follow in the Historic Center of

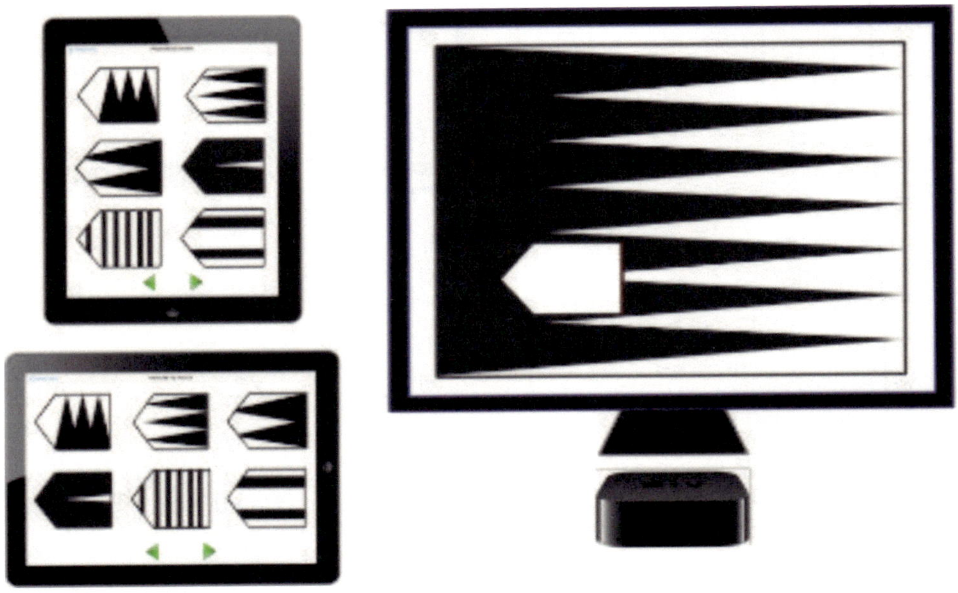

Figure 4.
DUI design for Raven test using the Tablet in landscape or portrait orientation with the Smart TV.

Mexico City to reach a point of interest around it, in a radius no greater than 5 km. The user can execute the application on the smartphone, on the smartwatch or in both devices using a DUI.

The user decides at any time to activate the DUI, from the smartphone or from the smartwatch. If the user activates the DUI from the smartphone and the smartwatch does not have the application active, then the application is activated and the smartphone sends the status of the application, which indicates that it is in some search of a site of interest or that it is displaying geographic information. In this case, the smartphone starts the application. If the user activates the DUI from the smatwatch and the application is not active on the smartphone, then the status of the application is activated and transferred, so that the smartphone knows the activity that it must present on the display.

Figure 5 shows the DUI for the search and guide application of sites of interest. The DUI uses the deployment area of the smartphone and smartwatch. In the case of search by predetermined sites, a list is presented on both devices. To guide the user to the site of interest, the smartphone presents the route on a map and an arrow indicating where to go. To guide the user to a site of interest, the smartwatch presents an arrow indicating the orientation.

Portability: Depending on the state of the application the UI is duplicated in both devices (for example, search for interest sites) or part of the UI is displayed on the smartwatch and the UI complete is displayed on smartphone.

Fragmentation: Several screens of the application are duplicated in the smartphone and in the smartwatch, for these cases there is not fragmentation. When the application displays the map to indicate the route to the user to reach a site of interest, the UI is fragmented into two elements: one element E1 displays the map and the element E2 displays the date indicating the orientation of the site of interest. The smartphone displays E1 and E2 and the smartwatch only displays E2.

Simultaneity: Both devices display in real time the changes made by the user.

Continuity: The user defines when starting using the DUI and when finishing. When the user decides to activate the DUI and the application in smartphone is synchronized with the application in the smartwatch.

3.3 Tablets-smartphones

This example uses a set of tablets or smartphones to increase the working area. The DUI is increased dynamically when the application detects another device with the application. The application is a mental maps editor. The application detects

Figure 5.
DUI design for an application using smartphone and smartwatch.

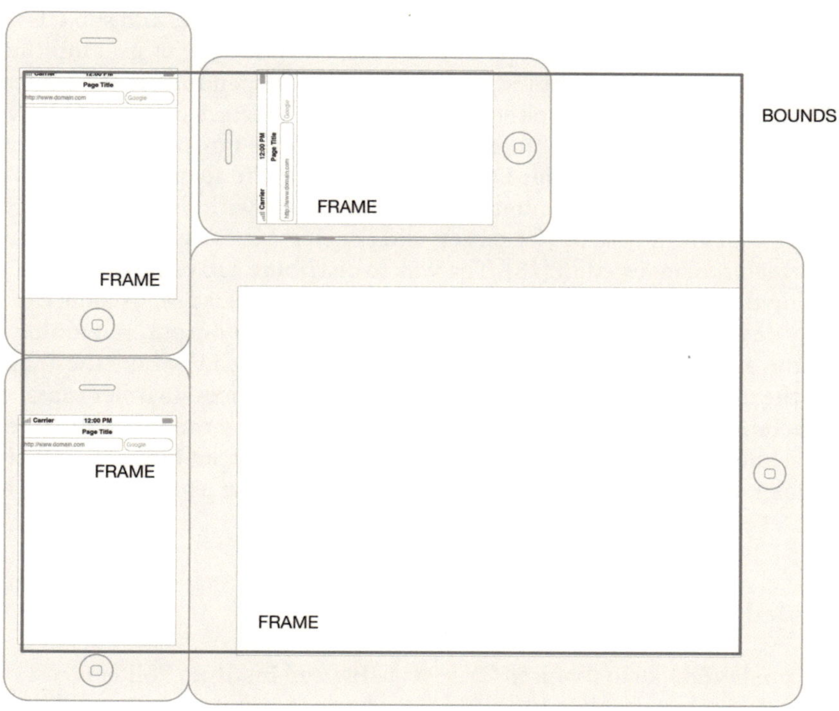

Figure 6.
DUI design for application using smartphones and tablets.

a gesture to add or remove a device from the application, and therefore adjust the DUI dynamically (**Figure 6**).

Portability: The UI has one element. This element is a canvas to draw and mental map. When a device is incorporated to the DUI, the element is duplicated in another device.

Fragmentation: Every display in the DUI display a part of general working area. Each device displays a part of the canvas. The canvas has a general work area called Bounds. And every device display a subarea called Frame. The Bounds is increases as devices are added.

Simultaneity: All devices must handle the changes in the Bounds. In addition to adding elements to the editor, this change of state of the canvas is sent to all the devices in the array. All devices must handle the changes in the Bounds. In addition to adding elements to the editor, this change of state of the canvas is sent to all the devices in the array. Thus, adding, removing or modifying canvas elements generates a message sending to the devices to update the state of the objects that are drawn on the canvas.

Continuity: When devices are added to the array, the canvas state is transferred to the new device and display one part of canvas.

4. Conclusions

The trends in DUIs its about the real time system for make the distributions; have software engineering methodologies for the design and implementation of DUIs; have consistent development frameworks and effectively incorporate the context management of applications and users. In this chapter the concept of mobile distributed user interface was made. This concept is based on models of distributed users interfaces, plasticity of user interfaces and mobile applications concepts. The concepts for DUIs help to indicate about elements and sub UIs that can be distributed and define the platforms host. The plasticity of user interface indicates when the applications must fragment de UI, depending on the context state. Now, the main issue in mobile distributed user interface is to decide how to adapt efficiently the sub user interfaces on its platform target. In this work we present some examples of mobile DUIs. We notice that the adaptability of sub user interfaces depends on the user interaction requirements with the application, that include the user group, the device target, and the elements of sub user interfaces, among others, as suggested in [28]. The way to distribute sub user interfaces depends on the application and the devices considers in their use. In some cases it is necessary duplicate elements to others devices but only for output, remaining the input in the original devices, such as the case Tablet-Smart TV, where the Tablet remains the user input, but the output is reply in both devices. In other cases, de GUI is decomposed and then, the input elements remain on the Tablet and the outputs elements are sent to Smart TV. In these cases the input interaction is always on the tablet. For the case of the tourist guide using a smart phone and smart watch, consider several scenarios depending on the user cases.

Acknowledgements

Authors should like to thank to Cinvestav-IPN and Instituto Politécnico Nacional, SIP project number 20196705 “Diseño de la arquitectura de middleware para protocolos criptográficos en dispositivos restringidos” by the resources provided and the facilities for this work.

Author details

Erika Hernández-Rubio[1*], Amilcar Meneses-Viveros[2] and Sonia G. Mendoza-Chapa[2]

1 Instituto Politécnico Nacional, SEPI-ESCOM, Mexico City, Mexico

2 Computer Science Department, Cinvestav-IPN, Mexico City, Mexico

*Address all correspondence to: ehernandezru@ipn.mx

IntechOpen

© 2019 The Author(s). Licensee IntechOpen. This chapter is distributed under the terms of the Creative Commons Attribution License (http://creativecommons.org/licenses/by/3.0), which permits unrestricted use, distribution, and reproduction in any medium, provided the original work is properly cited.

References

[1] Rashedul I, Islam R, Mazumder T. Mobile application and its global impact. International Journal of Engineering & Technology (IJEST). 2010;**10**:6:72-78

[2] Tesoriero R, Gallud JA, Lozano MD, Penichet VM, Vanderdonckt J. Distributed user interfaces: Collaboration and usability. In: CHI'12 Extended Abstract on Human Factors in Computing Systems; 5-10 May 2012; Austin, Texas: ACM; 2012. pp. 2719-2722

[3] Villanueva PG, Tesoriero R, Gallud JA. Is the quality in use model valid for distributed user interfaces. In: Proceedings of the 2nd Workshop on Distributed User Interfaces: Collaboration and Usability (DUI 2012); 5-10 May 2012; Austin, Texas: ACM; 2012. pp. 39-44

[4] Cutugno F, Leano VA, Rinaldi R, Mignini G. Multimodal framework for mobile interaction. In: Proceedings of the International Working Conference on Advanced Visual Interfaces; 21-25 May; Capri Island, Italy: ACM; 2012. pp. 197-203

[5] Raptis D, Tselios N, Kjeldskov J, Skov MB. Does size matter?: Investigating the impact of mobile phone screen size on users' perceived usability, effectiveness and efficiency. In: Proceedings of the 15th International Conference on Human-Computer Interaction with Mobile Devices and Services. ACM; 2013. pp. 127-136

[6] Leite FLV, Prietch SS, Preti JPD. Empowerment of assistive technologies with mobile devices in a DUI ecosystem. Procedia Computer Science. 2015;**67**:358-365

[7] Caballero PC et al. Distributed user interfaces for Poppelreuters and Raven visual tests. In: International Conference on Human Aspects of IT for the Aged Population. Cham: Springer; 2017. pp. 325-338

[8] Luyten K, Coninx K. Distributed user interface elements to support smart interaction spaces. In: Seventh IEEE International Symposium on Multimedia (ISM'05); IEEE; 2005. p. 8

[9] Gallud JA, Penichet VMR. Distributed user interfaces: Distributing interactions to facilitate universal access. Universal Access in the Information Society. 2017:1-2

[10] Tesoreiro R, Altalhi AH. Model-based development of distributable user interfaces. Universal Access in the Information Society. 2017:1-28

[11] Sanctrorum A, Signer B. Towards end-user development of distributed user interfaces. Universal Access in the Information Society. 2017:1-15

[12] Vanderdonckt J. Distributed user interfaces: How to distribute user interface elements across users, platforms, and environments. In: Proceedings of XI Interacción; 2010. pp. 20

[13] Sjölund M, Larsson A, Berlung E. Smartphone views: Building multi-device distributed user interface. In: Mobile HCI 2004; Springer; 2004. pp. 507-511

[14] Antonio P et al. Defining distribution constraints in distributed user interfaces. Journal of Universal Computer Science. 2013;**19**(6):831-850

[15] Melchior J, Vanderdonckt J, Van Roy P. A model-based approach for distributed user interfaces. In: Proceedings of the 3rd ACM SIGCHI symposium on Engineering interactive computing systems; ACM; 2011. pp. 11-20

[16] Melchior J, Vanderdonckt J, Van Roy P. Distribution primitives for distributed user interfaces. In: Distributed User Interfaces. London: Springer; 2011. pp. 23-31

[17] Gallud JA et al. A proposal to validate the user's goal in distributed user interfaces. International Journal of Human Computer Interaction. 2012;**28**(11):700-708

[18] Peñalvert A et al. Distributed user interfaces: Specification of essential properties. In: Distributed User Interfaces. London: Springer; 2011. pp. 13-21

[19] Hernandez IMT, Viveros AM, Rubio EH. Analysis for the design of open applications on mobile devices. In: Proceedings of CONIELECOMP 2013, 23rd International Conference on Electronics, Communications and Computing; IEEE; 2013. pp. 126-131

[20] Wasserman T. Software engineering issues for mobile application development. In: FoSER 2010; 2010

[21] Thevenin D, Coutaz J. Adaptation and plasticity of user interfaces. In: Workshop on Adaptive Design of Interactive Multimedia Presentations for Mobile Users; 1999. pp. 7-10

[22] Coutaz J, Calvary G. HCI and software engineering for user interface plasticity. Human-Computer Interaction Handbook: Fundamentals, Evolving Technologies, and Emerging Applications. 3rd ed. CRC Press; 2012. pp. 1195-1220

[23] Vanderdonckt J et al. Multimodality for plastic user interfaces: Models, methods, and principles. In: Multimodal User Interfaces. Berlin, Heidelberg: Springer; 2008. pp. 61-84

[24] Winograd T. Architectures for context. Human Computer Interaction. 2001;**16**(2-4):401-419

[25] Coutaz J. User interface plasticity: Model driven engineering to the limit. In: Engineering Interactive Computing Systems (EICS 2010) International Conference; Berlin Germany: ACM; 2010. pp. 1-8

[26] Sottet JS, Calvary G, Favre JM. Models at run-time for sustaining user interface plasticity. In: Models@ run. time workshop; 2006

[27] Sottet JS, Ganneau V, Calvary G, Coutaz J, Demeure A, Gavre JM, et al. Model-driven adaptation for plastic user interfaces. In: IFIP Conference on Human Computer Interaction; Heidelberg Berlin: Springer; 2007. pp. 397-410

[28] Sottet JS, Calvary G, Coutaz J, Favre JM. A model-driven engineering approach for the usability of plastic user interfaces. In: IFIP Conference on Human Computer Interaction; Heidelberg Berlin: Springer; 2008. pp. 140-157

[29] Calvary G, Coutaz J, Thevenin D. Supporting context changes for plastic user interfaces: A process and a mechanism. In: Blandford A, Vanderdonckt J, Gray P, editors. People and Computers XV—Interaction without Frontiers. London: Springer; 2001. pp. 349-363

[30] Thevenin D, Coutaz G. Plasticity of user interfaces: Framework and research agenda. In: Human-Computer Interaction (INTERACT'99). Edinburgh: IOS Press; 1999. pp. 110-117

Chapter 8

Metabolic Health Analysis and Forecasting with Mobile Computing

Zsolt P. Ori

Abstract

The goal of this paper is to demonstrate feasibility of a concept of mobile computing to help users to reach and maintain metabolic health. For this purpose, we analyze data from 12 clinical studies with a total of 39 study arms from the international literature to show that insulin resistance measured by HOMA-IR could be followed and its changes could be predicted using our weight-fat mass-energy balance calculations taking advantage of the significant and strong correlation between changes of HOMA-IR and state variables of the energy metabolism like changes of weight, fat mass, R-ratio, Rw-ratio, and fat burning fraction of the energy production. We introduce here our extended weight-fat mass-energy balance calculation to assess de novo lipogenesis, adaptive thermogenesis, and the 24-hour nonprotein respiratory quotient. We show how we can analyze and predict individualized state variables of the metabolism, which serve as metrics for the quantification of the interrelationship between energy metabolism and insulin resistance facilitating management and self-management of insulin-resistance related conditions including obesity, fatty liver, prediabetes, metabolic syndrome, and type 2 diabetes. The feedback of individualized metrics using tools of the digital health era may amount to channeling focus also to patient-centered individualized care and to accelerating nutrition research.

Keywords: energy metabolism, insulin resistance, metabolic monitoring, mobile computing, 24-h nonprotein respiratory quotient, fat oxidation, carbohydrate oxidation, de novo lipogenesis, adaptive thermogenesis, obesity, fatty liver, prediabetes, metabolic syndrome, type 2 diabetes, cardiovascular morbidity, mortality, dynamic changes of behavior, lifestyle modification, patient-centered individualized care, digital health

1. Introduction

The purpose of this paper is to outline a new proposed direction of managing and self-managing metabolic health including insulin resistance in the era of mobile technology.

I am perhaps a rare breed of internist with previous training in biomedical cybernetics before entering medical school. My training and research experience in control engineering together with my experiences as a clinician in academic and nonacademic settings have inspired me to use mathematical modeling tools to tackle the most sweeping health problem of our time which has reached now pandemic proportion all over the world, i.e., insulin resistance with its devastation in terms of rising cardiovascular disease (CVD), morbidity, and mortality. The reason for focusing on insulin resistance is the overwhelming evidence that insulin resistance syndrome has been proven to be an independent risk factor for CVD mortality, and effective, clinically usable indicators can be derived from readily measured variables [1].

Central to the mission of primary care is fighting the burden of noncommunicable chronic diseases including the most prominent one, CVD [2]. CVD is substantially higher in individuals with unhealthy lifestyle characteristics, including visceral obesity, prediabetes, diabetes, insulin resistance, metabolic syndrome, physical inactivity, poor diet, and cigarette smoking. One could ask the question, why is insulin resistance such an important issue and how could we address this problem more effectively? I see a four-pronged answer to this question:

1. There is a need for heightened awareness of the pathophysiological processes at play in CVD which is accelerated inflammation leading to atherosclerosis driven by insulin resistance. The higher level than normal of inflammatory markers and cytokines are triggered by complex cell physiology changes taking place initially in the visceral adipose tissue and likely related to the regulation of the deposition of the newly synthetized fat. A recently published research article [3] found new insight into the pathophysiological steps of how insulin resistance with or without obesity begets proinflammatory changes [4, 5] occurring, among others, at the arterial walls.

2. Measuring insulin resistance in clinical practice is a huge challenge. The gold standard is the "hyperinsulinemic euglycemic clamp," which measures the amount of glucose necessary to compensate for an increased insulin level without causing hypoglycemia. Nowadays the test is rarely performed in clinical care. The frequently used clinical assessment is with the homeostatic model assessment of insulin resistance (HOMA-IR) which closely mirrors the glucose clamp technique [6]. This requires only point-of-care invasive measurement of fasting insulin and glucose level. There is a need for a continuous noninvasive measurement method which can provide the same information as HOMA-IR.

3. A lifelong heightened awareness is needed in our accelerated world which pulls us back to optimum decision-making by mindfulness regarding eating and exercising. We have biased perceptions regarding how much we eat [7], and we possess no bodily sensation regarding the size of the visceral fat (a prime source of insulin resistance) and its daily changes. The privacy of a personalized gage such as a smart watch or phone app is needed to gage changes of our metabolic health and fitness level [8].

4. For tracking insulin resistance noninvasively, there is a need to know the total fat balance [9–11] which includes also the daily de novo lipogenesis (DNL). Currently the standard way to measure DNL is the 24-h metabolic chamber [12, 13]. Mobile computing technology just may offer a key solution to this issue.

The tools needed to realize the new proposed approach are embedded into the science of cybernetics. Cybernetics is mostly concerned with exploring regulatory systems—their structures, constraints, and possibilities using mathematical modeling. Cyber-therapy is defined here as the combined use of individualized mathematical-statistical modeling, prediction, planning for change, and gaining control and self-management of the metabolism with the option for guided therapy by feedback of information on components of the human energy metabolism of an individual. Our mathematical modeling techniques of the human energy metabolism [8–11] is a tool to observe the difficult-to-measure variables of the human energy metabolism, such as slowly occurring body composition changes including lean mass, protein mass, fat mass, extracellular and intracellular water mass, or practically impossible-to-measure variables like utilized macronutrient intake, oxidation rates, and the de novo lipogenesis. The overall goal of individualized cyber-therapy is to gain better control of the metabolism, i.e., reach and maintain optimum body composition and cardiometabolic fitness with minimized effort over the shortest possible time while staying well hydrated and maintaining optimal insulin sensitivity through self-management with mindfulness and/or guided therapy.

My recent paper to the same publisher [8] introduces a proposed cyber-physical system (CPS) as a framework to manage and self-manage metabolic health including insulin resistance. The essential elements of CPS comprise smart watch with appropriate sensors, smart phone, and bathroom scale with fat weight measuring capabilities. The various devices with their apps are connected through cloud computing. The main software component is a metabolic health monitoring app (MHM) performing data gathering and result display of metabolic trends [11]. MHM can make predictions regarding changes of the metabolic state variables (SVs) such as fat mass, lean body mass, insulin resistance changes by the Rw-ratio, 24-h nonprotein respiratory quotient, as well as the utilized macronutrient intake and oxidation rates. We developed mathematical models of the human energy metabolism allowing for estimation of the SVs [9–11] requiring serial fat weight and lean body mass measurements [9, 10]. We introduced our weight-fat weight-energy balance (WFE) calculations requiring only serial weight and fat weight measurements for basic calculations to estimate changes of insulin resistance [8]. In the same paper, we provided also evidence for feasibility of the CPS concept in healthy young men to track and predict insulin resistance.

Central to the goal of providing metrics for the quantification of insulin resistance is the recognition of its interrelationship with other easily and daily measurable state variables like weight and fat mass. In this regard we take advantage of the observation that there is a correlation between BMI/weight/body composition and insulin resistance measured, for example, with HOMA-IR [14, 15, 27]. We reported earlier that we found significant negative correlation between HOMA-IR and R-ratio or Rw-ratio [8, 10, 11, 15]. It has been also our research hypothesis that one could exploit the strong inverse correlation between HOMA-IR and the R-ratio or Rw-ratio and use this to measure indirectly changes of insulin resistance derived from serial weight and fat weight measurements [8].

The goal of this paper is to contribute to the developing field of mobile technologies and their use for health-related applications in three areas:

1. More evidence is provided for the connection between HOMA-IR and WFE calculations for clinical practice: here we show feasibility of our research hypothesis that insulin resistance changes by HOMA-IR can be predicted by using the WFE calculation framework in a wide variety of clinical scenarios involving insulin resistance changes and not just in young healthy men as it was already demonstrated in [8]. For this purpose, we use data from [16–27] to show further support for the idea that insulin resistance measured by HOMA-IR could be followed and its change could be predicted by WFE calculation.

2. We give here our theoretical considerations on how DNL and adaptive thermogenesis (AT) could be assessed with mobile technology during acute phase and at predicted steady-state equilibrium of the energy metabolism. AT becomes important when the energy metabolism goes from one steady state to another and the AT energy production or disappearance will oppose the direction of change [28]. We would like to introduce here our extended weight-fat weight-energy balance calculation allowing for DNL and AT calculations (WFE-DNL-AT) and pointing out its limitations.

3. I give here supportive results to the proposition of using WFE-DNL-AT calculations and show how insulin resistance changes with WFE-DNL-AT calculations compared with results of indirect calorimetry obtained by 24-h measurement in a metabolic chamber. For all of these purposes, I reanalyze the published trial data of the "Calorie for Calorie, Dietary Fat Restriction Results in More Body Fat Loss than Carbohydrate Restriction in People with Obesity" (CC trial) [29] using WFE-DNL-AT calculations.

All mathematical tools of WFE-DNL-AT calculation are summarized in Appendix.

2. Method of the correlation analysis between HOMA-IR and chosen state variables

This meta-analysis utilizes our dynamic energy balance equation (Eq. (1)) as it was introduced to the reader in [8]. This establishes a weight, fat weight, and energy balance calculation (WFE) based on the following mass and energy relationship:

$$\varrho_{Wk} \cdot \Delta W_k + \varrho_F \cdot \Delta F_k = \left(\varrho_{Wk} \cdot Rw_k + \varrho_F\right) \cdot \Delta F_k = MEI_k - TEE_k = EB_k. \quad (1)$$

Essentially, the equation expresses the equivalent change of weight represented here as $\Delta W_k = Rw_k \cdot \Delta F_k$ and fat weight ΔF_k in response to the energy balance EB_k, i.e., the difference of metabolized energy intake MEI_k and total energy expenditure TEE_k on a given day k. We listed in the glossary the meaning of each variable.

Eqs. (2)–(18) constitute the framework derived from Eq. (1) for the correlation analysis between the percentage change of HOMA-IR $\Delta H\%$ and changes of state variables Rw-ratio ΔRw, weight ΔW, and fat burning fraction $\Delta\chi$ over the course of a clinical trial. We used 39 study arms from 12 clinical trials with a variety of length of the studies performed ranging from 3 days to 365 days [16–27]. The correlation analysis was done in MATLAB. The outcome results of the trials are shown in **Table 1**. Here n is the number of days in the clinical trial, *subjects* means the number of participants, ΔW_{n-0} designates the average weight change in kilograms during the trial period, ΔF_{n-0} symbolizes the average fat weight change during the trial period in kilograms, and ΔH_{n-0} stands for the average change of HOMA-IR with sugar in mg/dL and insulin in mU/L.

I calculate the average lean mass change ΔL_{n-0} as the difference between average weight change and fat weight change as in Eq. (2):

$$\Delta L_{n-0} = \Delta W_{n-0} - \Delta F_{n-0}. \quad (2)$$

For current calculations we assume that the energy density value of lean mass change ϱ_{Lk} will remain stable, and it takes the value around $\varrho_L \approx 1.8$ kcal/g which is a value quoted in the literature [30], but its real value is unknown and uncertain. Likewise, the energy density parameter for weight change ϱ_{Wk} is calculated using the energy density value of lean mass change $\varrho_L \approx 1.8$ kcal/g and using the energy relationship as in Eq. (3):

$$\varrho_{Wk} = \varrho_L \cdot \frac{\Delta L_k}{\Delta W_k}. \tag{3}$$

The αw_k first-order term coefficient of the weight-fat logarithmic relationship is thought to be stable during the trial period under the stationarity assumption and therefore remains unindexed denoted as αw . Its value is calculated from weight on the first day $k = 0$ and last day $k = n$ of the study as in Eq. (4):

$$\alpha w = \frac{W_n - W_0}{ln\ F_n - ln\ F_0}. \tag{4}$$

The weight change and fat weight change on the first day and last day is calculated through several steps as in Eqs. (5)–(8):

Estimated daily energy balance for each day is the same EB_0 as in Eq. (5):

$$EB_0 = \varrho_L \cdot \frac{(\Delta W_{n-0} - \Delta F_{n-0})}{n} + \varrho_F^* \cdot \frac{\Delta F_{n-0}}{n} \tag{5}$$

The first day's fat weight change ΔF_1 and F_1 is calculated as in Eqs. (6) and (7):

$$\Delta F_1 = EB_0 \cdot \left(\varrho_W \cdot \frac{\alpha w}{F_0} + \varrho_F^* \right)^{-1} \tag{6}$$

$$F_1 = F_0 + \Delta F_1 \tag{7}$$

Here ϱ_F^* can take different values: for net fat loss $\varrho_F^* \approx \varrho_F$, it takes the value of 9.4 kcal/g; for net fat synthesis, the value is 9.4 + 2.38 kcal/g because synthesis cost of fat from glucose is added to the energy density of fat.

The last day's fat weight change ΔF_{n-1} and F_{n-1} is calculated as in Eqs. (8) and (9).

$$\Delta F_{n-1} = EB_0 \cdot \left(\varrho_W \cdot \frac{\alpha w}{F_n} + \varrho_F^* \right)^{-1} \tag{8}$$

$$F_{n-1} = F_n - \Delta F_{n-1} \tag{9}$$

The first day's and last day's weight changes are calculated as in Eqs. (10) and (11):

$$\Delta W_1 = \alpha w \cdot (\ ln\ F_1 - \ ln\ F_0) \tag{10}$$

$$\Delta W_{n-1} = \alpha w \cdot (\ ln\ F_n - \ ln\ F_{n-1}) \tag{11}$$

The first day's and last day's Rw-ratio Rw_1 and Rw_{n-1} is calculated as in Eqs. (12) and (13).

$$Rw_1 = \frac{\Delta W_1}{\Delta F_1} \tag{12}$$

$$Rw_{n-1} = \frac{\Delta W_{n-1}}{\Delta F_{n-1}} \tag{13}$$

Study	*n days*	*Subjects*	ΔW_{n-0}	ΔF_{n-0}	ΔH_{n-0}
[16] Hypoenergetic	84	44	−7	−4.8	−0.12
[16] Hypoenergetic + walking	84	38	−8.8	−6.5	−0.56
[17] Low-carbohydrate	56	12	−7.4	−3.9	−1.3
[17] Low-fat	56	12	−6.5	−3.8	−0.6
[18] Healthy low-fat	365	305	−5.3	−3.7	−0.8
[18] Healthy low-carb	365	304	−6	−4.1	−0.7
[19] Hypo_LF/HC	5	10	−1.1	−1.2	−0.9
[19] Hypo_HF/LC	5	8	−1.5	−1.5	−0.9
[20] Whey protein 4 days	4	8	−5.4	−2.2	−1.8
[20] Sucrose 4 days	4	7	−4.3	−1.9	−1.1
[20] Whey protein 4 + 3 days	3	8	1.3	−0.6	0.6
[20] Sucrose 4 + 3 days	3	7	0.8	−0.9	0.4
[20] Whey protein 7 + 28 days	4	8	−0.5	−0.8	1.3
[20] Sucrose 7 + 28 days	4	7	−0.6	−1.3	0.8
[21] Low-glycemic 180	180	32	−4.5	−2.4	−0.5
[21] Low-fat 180	180	34	−3.75	−1.4	−0.2
[21] Low-glycemic 18-mo	360	32	1.75	0.8	0.4
[21] Low-fat 18-mo	360	34	2.25	0.3	0.2
[22] Women obese	77	17	−12.2	−9.5	−1.7
[22] Men obese	77	17	−17.6	−12.3	−1.8
[23] Women training 1	168	18	−2.7	−2.2	−0.1
[23] Women training 2	168	16	−2.3	−1.8	−0.2
[24] HC 2 day	2	11	−1.61	−0.20	−1.11
[24] LC 2 day	2	11	−2.64	−0.04	−1.69
[24] HC 77 day	77	11	−5.75	−4.17	−0.15
[24] LC 77 day	77	11	−5.09	−5.22	−0.15
[25] A_calorie Restriction	7	10	−3	−2.2	−0.61
[25] A_refeeding	7	10	3.1	1.4	0.82
[25] B_overfeeding	7	10	1.6	0.7	0.84
[25] B_calorie restriction	7	10	−3.4	−1.9	−1.08
[26] Women control M	365	87	−0.42	−0.11	−0.03
[26] Women control O	365	87	−1.34	−1.71	−0.12
[26] Women exercise M	365	117	−1.67	−3.08	−0.43
[26] Women exercise O	365	117	−2.67	−1.42	0.15
[26] Women diet M	365	118	−7.64	−11.15	−0.73
[26] Women diet O	365	118	−6.04	−5.25	−0.82
[26] Women diet+exercise M	365	116	−8.59	12.95	−0.75
[26] Women diet+exercise O	365	116	−9.58	−8.99	−0.88
[27] Correlation IS and WL	336	72	−10.8	−8.3	−14.2

Table 1.
Outcome results of 12 clinical trials with a total of 39 study arms.

We calculate the absolute change of the Rw-ratio ΔRw over duration of the trial as in Eq. (14):

$$\Delta Rw = Rw_{n-1} - Rw_1 \tag{14}$$

We calculated the fat burning fraction using Eqs. (15) and (16):

$$\chi_1 = \frac{\varrho_{W1} \cdot Rw_1}{\varrho_{W1} \cdot Rw_1 + \varrho_F} \tag{15}$$

$$\chi_{n-1} = \frac{\varrho_{W_{n-1}} \cdot Rw_{n-1}}{\varrho_{W_{n-1}} \cdot Rw_{n-1} + \varrho_F} \tag{16}$$

We define the absolute change of fat oxidation fraction $\Delta\chi$ as in Eq. (17):

$$\Delta\chi = \chi_{n-1} - \chi_1 \tag{17}$$

We calculated the percent change $\Delta H\%$ of the HOMA-IR over the duration of the trial as in Eq. (18) where H_0 is the average HOMA-IR value at baseline:

$$\Delta H\% = \frac{\Delta H_{n-0}}{0.01 * H_0} \tag{18}$$

3. Method of the analysis of the CC trial data with WFE-DNL-AT calculations

The detailed description of the WFE calculation with the capability of calculating DNL and adaptive thermogenesis/thermal loss (WFE-DNL-AT) is detailed in Appendix A. For demonstration purpose we apply this method to analyze the published result of the CC trial [29]. The CC trial [29] investigated 19 adults with obesity. The intervention was the selective dietary restriction of carbohydrate (RC) versus fat (RF) for 6 days following a 5-day baseline diet. Subjects received both the isocaloric baseline diets followed by either RC or RF diet in random sequence during two inpatient stays when they were confined to a metabolic ward for two 2-week periods each. The 24-h nonprotein respiratory quotient Rnp'_k was measured on day −4, 0, 1, 4, and 6 in a 24-h respiratory chamber. The baseline measurements were taken at day 0. The published data of the CC trial were sparse, and the results of laboratory measurements were published only for baseline and for the end point. I generated the needed daily weight W_k and fat weight F_k data using MATLAB's interpolation function 'pchip.' First, I calculated the daily Rw-ratio Rw_k, fat burning fraction χ_k, 24-hour nonprotein respiratory quotient Rnp_k, de novo lipogenesis DNL_k, and adaptive thermogenesis T_k without the a priori knowledge of the measured Rnp'_k on day 1, 4, and 6 using the uncorrected WFE-DNL-AT algorithm as in (A1, A2, A3, A6, A7). Second, I performed inverse calculations using the uncorrected WFE-DNL-AT model and the measured nonprotein respiratory quotient Rnp'_k on days 1, 4, and 6 as input to arrive at the indirectly measured de novo lipogenesis mDNL and adaptive thermogenesis mT denoted mDNLRC, mTRC for the RC arm and mDNLRF, mTRF for the RF arm of the CC trial. Finally, I analyzed how the fat intake fraction φ_k could be used to predict better the measured nonprotein respiratory quotient Rnp'_k and to estimate mDNLRC, mTRC, mDNLRF, and mTRF and how to build a corrected WFE-DNL-AT, which includes (A4) for RF or (A5) for RC diet and could work without the a priori knowledge of the measured 24-h respiratory quotient Rnp'_k.

The steps of calculations followed the WFE algorithm as in [8], and for DNL and AT calculations, I used Eqs. (A1)–(A7). The unknown values for the adaptive thermogenesis coefficient c_{AT} and the unknown fraction of the metabolized carbohydrate intake for de novo lipogenesis c_{DNL} were assumed to be time independent for this analysis and their values were estimated with the assumption that at baseline the energy balance is zero and the energy system is at steady state. I used the same recursive minimization procedure as already explained in [8] to calculate αw_k, the first-order term coefficient of the weight-fat logarithmic relationship, and the energy density for weight ϱ_{Wk} and the Rw-ratio Rw_k.

4. Results of the correlation analysis between HOMA-IR and chosen state variables

The results of correlation analysis across 12 clinical trials with a total of 39 study arms are summarized in **Table 2**. The percent change $\Delta H\%$ of the HOMA-IR over the duration of the trials was correlated with the absolute change ΔRw of the Rw-ratio, absolute weight change ΔW, change of fat oxidation fraction $\Delta\chi$, and the absolute fat mass change ΔF.

A sub-analysis was also performed with the results of correlation analysis of three clinical trials, [18, 26, 27], with inclusion of 11 study arms. The rationale for this sub-analysis was that all of them were long-term studies with duration of equal or longer than 336 days with satisfying the stationarity requirement for the analysis. The results of the sub-analysis are in **Table 3**.

	$\Delta H\%$	P value
ΔRw	−0.6745	0.0000024
ΔW	0.6413	0.0000108
$\Delta\chi$	0.6218	0.0000238
ΔF	0.4748	0.0022542

Table 2.
Correlation results of 12 clinical trials with a total of 39 study arms.

	$\Delta H\%$	P value
ΔRw	−0.8481	0.0009699
ΔW	0.8890	0.0002512
$\Delta\chi$	0.8206	0.0019656
ΔF	0.7605	0.0065810

Table 3.
Correlation results of three clinical trials: [18, 26, 27] with a total of 11 study arms.

5. Results of the analysis of the CC trial data with WFE-DNL-AT calculations

The results of the WFE-DNL-AT calculations using published data of the CC trial [29] are shown in **Figures 1–6**.

In **Figure 1** the measurement points mFRC and mFRF and trajectories of the fat mass change in the RC and RF arm of the CC study are shown. The dashed lines FRC and FRF are the results of the WFE calculation.

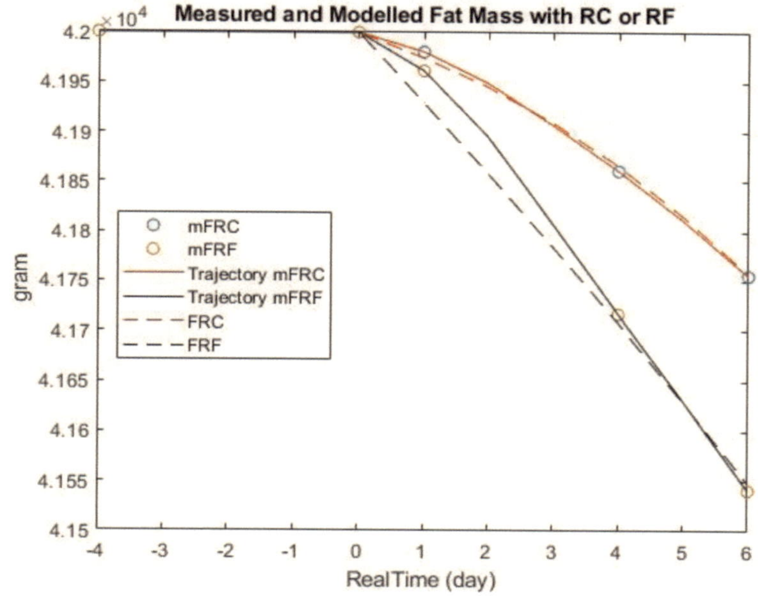

Figure 1.
Measured and model calculated loss of fat mass without apriori knowledge of measured Rnp'_k

In **Figure 2** the measurement points mWRC and mWRF and trajectories of body weight change in the RC and RF arm of the CC study are depicted. The dashed lines WRC and WRF are the results of the WFE calculation.

In **Figure 3** the measurement points and trajectories of HOMA-IR are shown in the RC and RF arm of the CC trial. The dashed lines are the model predicted calculations of the fat burning fraction χ mRC and χmRF, and the dotted line represents the Rw-ratios RwRC and RwRF in the RC and RF arm of the CC study.

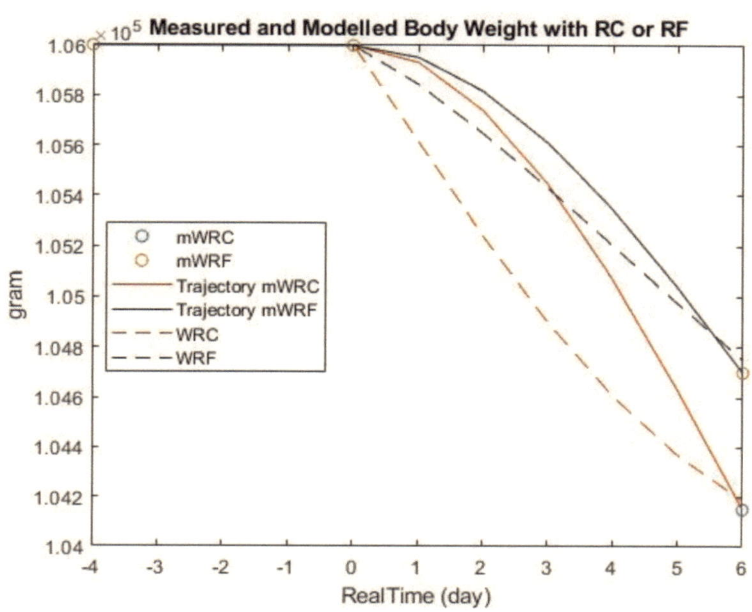

Figure 2.
Measured and model calculated weight loss without a priori knowledge of measured Rnp'_k.

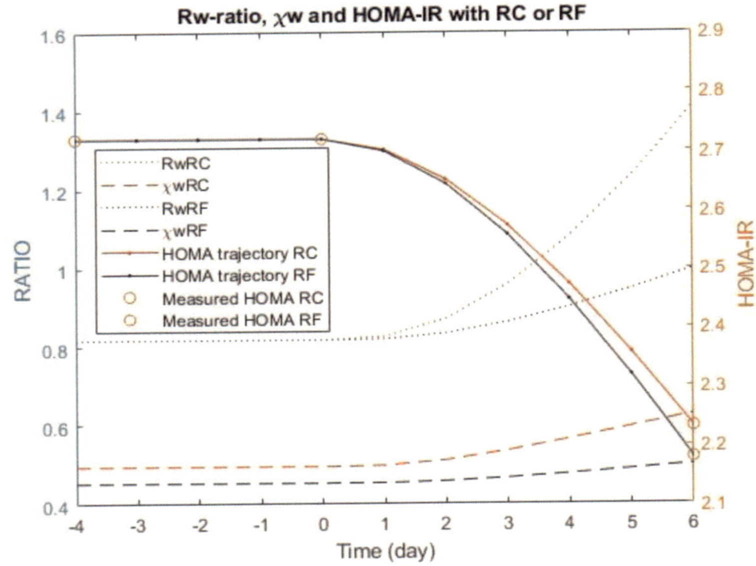

Figure 3.
Model calculated Rw-ratio, fat burning fraction χ_k without a priori knowledge of measured Rnp_k'.

The measurement points and trajectories of the measured nonprotein respiratory quotient Rnp_k' in the RC and RF arm of the CC study are depicted as mRnpRC and mRnpRF in **Figure 4**. The dashed lines are model predicted calculations of the nonprotein respiratory quotient Rnp_k' by the uncorrected WFE-DNL-AT model labeled as model RnpRC and model RnpRF. The dotted lines denoted as corrected RnpRC and corrected RnpRF are results calculated by the corrected WFE-DNL-AT model.

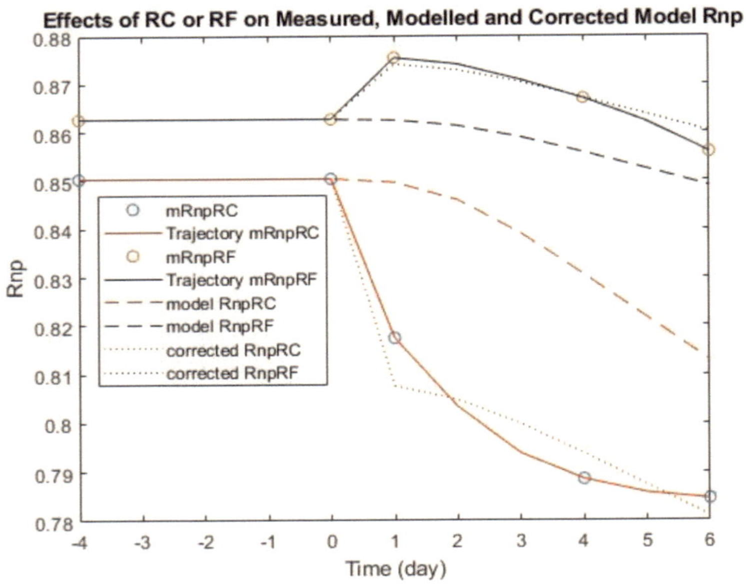

Figure 4.
Measured nonprotein respiratory quotient Rnp_k', uncorrected WFE-DNL-AT model predicted nonprotein respiratory quotient Rnp_k, and corrected WFE-DNL-AT model predicted nonprotein respiratory quotient Rnp_k.

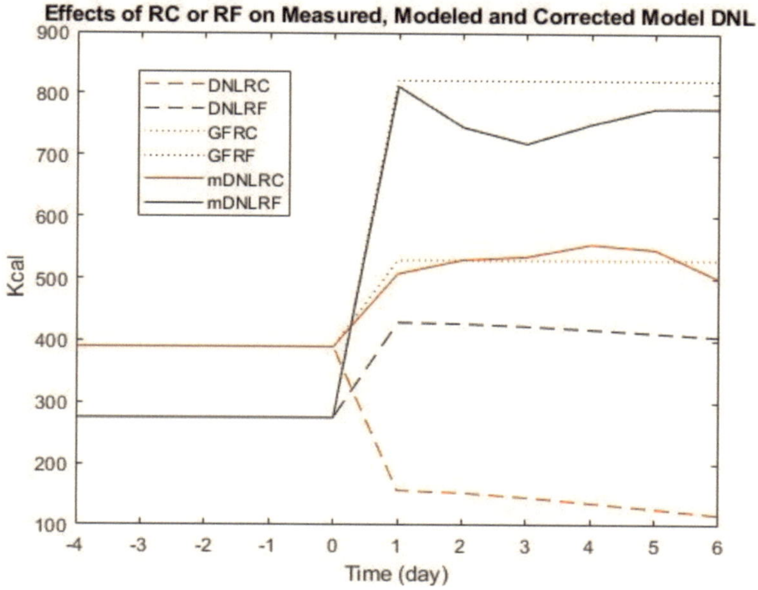

Figure 5.
The indirectly calculated DNL from measured nonprotein respiratory quotient Rnp'_k, DNL prediction by the uncorrected WFE-DNL-AT model, and predicted DNL by the corrected WFE-DNL-AT model calculation.

Figure 5 is to demonstrate the results of DNL calculations using measured Rnp'_k on day 1, 4, and 6 using inverse calculations with the uncorrected WFE-DNL-AT model in the RC and RF arm of the CC study marked as mDNLRC, and mDNLRF. The uncorrected WFE-DNL-AT model predicted results are denoted as DNLRC and DNLRF marked with dashed lines. The dotted lines GFRC and GFRF are showing the results of the corrected WFE-DNL-AT model calculations for DNL.

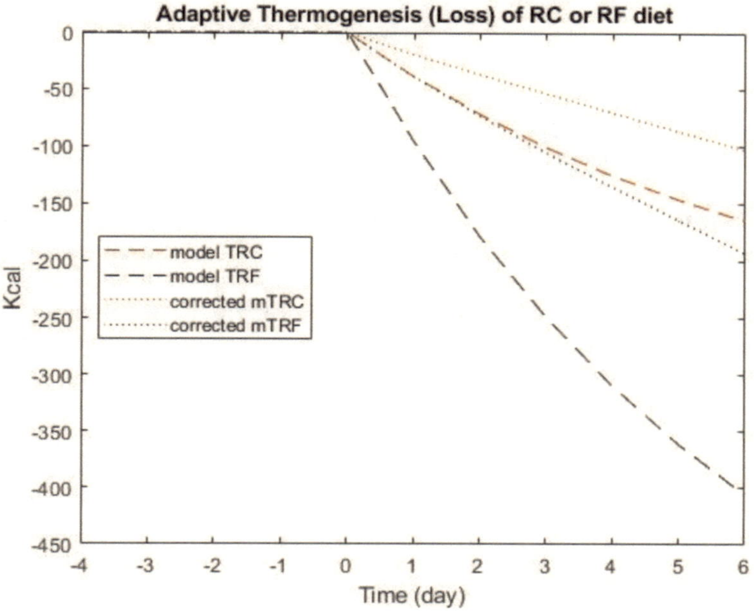

Figure 6.
The indirectly calculated adaptive thermogenesis using the uncorrected WFE-DNL-AT model and the corrected WFE-DNL-AT model calculation.

	RC diet		RF diet	
	$\Delta H\%$	P value	$\Delta H\%$	P value
Fat mass	0.9958845	0.9916711	1.0	0.9916711
Body weight	1.0	0.0000000	1.0	0.0000000
R-ratio	−0.9964298	0.0000000	−0.9973225	0.0000000
Rw-ratio	−0.9973225	0.0000000	−0.9979249	0.0000000

Table 4.
Correlation results in the CC [29] study.

	RC diet		RF diet	
	Mean	*Stand.dev.*	*Mean*	*Stand.dev.*
Error fat mass	−0.2921	3.2736	−0.1098	5.4886
Error weight	185.7393	233.7933	55.4621	80.9364
Error lean mass	178.7499	217.0219	15.8333	26.239

Table 5.
Error of prediction results in grams in the CC [29] study.

The adaptive thermogenesis/thermal loss calculation is in **Figure 6**. The dashed lines labeled with model TRC and model TRF are the results of the uncorrected WFE-DNL-AT model in the RC and RF arm of the CC study, respectively. The dotted lines labeled as corrected mTRC and corrected mTRF show the corrected WFE-DNL-AT results.

The correlation results between HOMA-IR and fat mass, body weight, R-ratio, and Rw-ratio are shown in **Table 4**. The errors of modeling fat mass, body weight, and lean mass in grams are shown in **Table 5**.

6. Discussion

The most important result of our meta-analysis across 12 clinical trials with a total of 39 study arms [16–27] is the high and significant correlation between changes of insulin resistance as measured with HOMA-IR and changes of Rw-ratio. The strength of this analysis is that the high correlation prevailed for all of the state variables examined regardless whether weight loss or weight gain was achieved during the trial and in the setting of a wide range of trial durations from 3 days to 365 days. As shown in **Table 2**, Rw-ratio ranked best regarding the level of correlation, followed by weight, fat weight, and fat burning fraction. Further, the high correlation is independent from dietary interventions such as isocaloric diet, overfeeding, or underfeeding or with or without exercise intervention. This result means also that the use of Rw-ratio as a surrogate marker for indirect measure of insulin resistance is justifiable for modeling changes of insulin resistance. The subanalysis looked at studies lasting longer than 336 days. The correlation coefficients scored as in **Table 3** are even higher than in **Table 2** with all the studies included. This predictive strength of the Rw-ratio regarding insulin resistance change could be even stronger in situations when strict steady-state energy balance is present as expected with longer study duration. An important advantage of the simple WFE analysis with Eqs. (1)–(18) is that only serial measurement of weight and fat weight is used and no calorie counting was done. The prevailing daily energy balance can be simply calculated as in Eqs. (2)–(5). Likewise, R-ratio or Rw-ratio can be obtained with simple arithmetic as in Eqs. (6)–(14), which can lead to the calculation of the fat burning fraction χ_k. Using Elia and Livesey's formula [31], the fat burning rate χ_k can be converted into the estimated 24-h nonprotein respiratory quotient Rnp_k. HOMA-IR change can be predicted by knowing that R-ratio and Rw-ratio are strongly and inversely correlated. This means that their reciprocal values could be used to predict proportional change of HOMA-IR. The strong correlation is in accordance with earlier findings of Thompson and Slezak [27] who were the first to report correlation between measures of insulin sensitivity and weight loss. They showed that the McAuley formula, which contains reference to triglyceride, showed greater correlation with weight loss than HOMA-IR, which does not contain information on lipids, supporting the idea that the sugar, insulin, and lipid kinetics and energy dynamics could and should be measured and modeled together. Therefore, extending the WFE model with the capability of estimating DNL like in the WFE-DNL-AT model is of theoretical as well as of practical importance for clinical use.

One weakness of the current analysis is that no individual data are published and the published data represent lumped together averages of weight and fat weight measurements at the beginning and at the end of the study period. The lack of individual serial data of weight or fat weight does not allow to have an exact insight into the dynamic of the process of body composition change which could occur along a concave or convex monotone decreasing function. Therefore, it is important to have the individual data and build an individual model which could be used also, among others, for interpolation to find missing data points.

Also, measuring insulin resistance with HOMA-IR has its own weaknesses, and maybe the McAuley formula could hold promise to improve model predictions.

The strength of the WFE modeling scheme with Eqs. (1)–(18) is that it makes minimal assumptions requiring only weight and fat weight data and it works also well for prediction of changes of HOMA-IR even when only the baseline value and the last measured value are available. The practicality of this matter is that the already commercially available measuring devices such as bioimpedance body composition analyzers are available, although their accuracy could be questioned. However, as a countermeasure we suggest the use of the Kalman filter [10, 11, 32, 33] minimizing the variance of the measurements and maximizing consistency.

The main contribution to science of the CC trial is that it offers an important glimpse into the acute phase reaction of the body's adaptation to the energy deficit state of glucose vs. fat. Even though RC diet increased the measured net fat disappearance more than the RF diet, the RF diet was more effective in the overall fat weight loss. I interpret this result as the body's physiological adaptation to a new and negative energy balance by increasing the production of readily usable needed fuel such as triglyceride or extra DNL_k^A as a way to meet demand above and beyond the predicted new steady-state level of DNL_1^B. The increased demand comes from the sudden drop of energy intake and ensuing energy deficit accompanied by relatively undisturbed carbohydrate and fat fuel burning rates of the body. The suddenly needed extra energy for DNL_1^A comes from different sources in the RC vs. RF dietary interventions. In RC with drop of available glucose, the needed energies come from the fat pool. In the RF scenario, the body readily grabs the available glucose at hand coming from undisturbed carbohydrate energy intake. While equal calorie deficit is created under both RC and RF, the RF state would sink more calories into new fat synthesis, i.e., DNL_k^A, as the fat synthesis from carbohydrate is an energy consuming process. This is because, according to Simonson [13], during lipogenesis each gram of lipid synthetized from glucose consumes $\varrho_{DNL} = 2.32$ kcal/g energy, and the process appears to be an energy "sink." In the RC diet, the needed extra triglyceride can be produced with simple lipolysis with no significant extra energy consumption or energy "disappearance or sink." Our modeling of this phenomenon with simple energy calculations for $GFRF_k$ as in Eq. (A4) and for $GFRC_k$ in Eq. (A5) is fully consistent with the measured result as demonstrated in **Figure 5**, where $GFRF_k$ and $GFRC_k$ and the measured DNL mDNLRC and mDNLRF are convincingly close to each other. The ratio of energy content of $GFRF_k$ and $GFRC_k$ reflects the difference of adaptation in RF vs. RC diet explainable now with Simonson's postulate for lipid synthesis from glucose [13] and using the energy constant $\varrho_{DNL} = 2.32$ kcal/g (see also Appendix A for more detail).

The strength of WFE-DNL-AT modeling is that the fat mass change predictions are accurate as in **Figure 1**, even under dietary interventions such as the RC or RF diet in the CC trial considered extreme. As fat mass measurement is a primary input data to our model calculations, its accuracy is essential to arrive at desired precision. Hall et al. [29] used DXA scan in the "CC" trial. Even with this very expensive tool, they felt that DXA still was inaccurate to determine fat mass in situations of dynamic weight change and shifting body fluids. Using commercially available bioimpedance scales for fat mass measurement with a one point in time, measurement certainly has much higher inaccuracy than DXA in determining the fat mass. However, currently most bioimpedance scales utilize the 50 kHz measuring frequency which gives quite suitable accuracy for extracellular water mass and intracellular water mass which resides mainly in lean mass, and the water content of the fat cell is negligible. In a way, the fat weight measurement by bioimpedance measurement is indirect: weight minus lean mass. This could be advantageous during weight loss when the fluids are shifting. Daily measurements with bioimpedance scale have the advantage that through obtaining a series of measurements, important secondary information related to weight, lean mass, and fat mass change can be obtained, and the variance of these measurements will allow for presentation of data in the $\pm$ standard deviation form. In response to the need for practicality, accuracy, and transparency in bioimpedance measurements, our company, Ori Diagnostic Instruments, LLC, has invented a Body Composition and Hydration Status Analyzer in unison with a high-frequency dielectric property analyzer [32, 33] which can quantify the size of intracellular and extracellular water along with fat mass better than commercially available bioimpedance analyzers because it measures the dielectric properties of tissue also at high frequency which is more suitable for fat mass change measurements. We take advantage of serial measurements by providing a posteriori values to a Kalman filter where the a priori estimates are obtained from our self-adaptive model of the human energy metabolism [9]. The combination of process model and measurement model equations combined with Kalman filter realizes a classic state-space modeling scheme [9], keeping the variance of measurements at minimum and maximizing consistency.

Regarding the body weight prediction with WFE-DNL-AT, far more challenges could be raised. In **Figure 2**, it is shown how the measured and predicted weight deviates, especially at day 3. Here are a high number of influencing factors at work. Probably most importantly, the daily measured weight decreases could follow an exponentially declining concave function rather than a convex function such as the applied interpolation function "pchip" in MATLAB. Beyond this modeling error, the lack of modeling of the extracellular and intracellular water mass change along with modeling glycogen and protein store changes is an issue in the current form of WFE-DNL-AT. In this regard our company, Ori Diagnostic Instruments, LLC, has created a patented modeling solution which includes also modeling of the intracellular as well as extracellular fluid volume changes and predicts also protein store as well as glycogen store changes [32, 33]. All results appear in the scientific form $\pm$ standard deviation, letting the user know about the accuracy and its change.

This demonstration study using CC trial data showed again the high level of correlation between HOMA-IR and R-ratio, Rw-ratio, weight, and fat mass as shown in **Table 4**. The modeling error of weight lean mass and fat mass was low as in **Table 5**. The calculated results for χ_k depicted as dashed lines in **Figure 3** are nicely pointing to the measured HOMA-IR values, demonstrating the predictive power of the calculated burning rate regarding HOMA-IR change prediction. This is driving home the main point that indirect measurement and prediction of HOMA-IR is possible, and the WFE model predicts that the RF diet leads to more reduction of insulin resistance and concomitant fat reduction than RC as measured.

In terms of modeling and predicting correctly changes of Rnp, the CC trial data were instrumental to improve our WFE-DNL-AT. It turns out that the truly measured Rnp'_k can be reproduced accurately at known calorie imbalance either with RC or RF diet as demonstrated in **Figure 4**. In the case of RC diet, the drop of glucose supply leads to lipolysis as in the model of $GFRC_k$ (see Appendix Eq. (A5)). During RF diet $GFRF_k$, energy will come mainly from available glucose and is easily quantifiable with our model (see Appendix Eq. (A4)). It has to be emphasized that the use of WFE-DNL-AT is possible only if the correct daily macronutrient energy of food is known.

This paper confirms for clinicians the long felt close relationship of insulin resistance to the energy metabolism, opening the theoretical opportunity to merge quantitative insulin sensitivity assessments such as the homeostasis model assessment [37] of glucose and insulin kinetics with quantitative modeling and measuring dynamics of the lipid metabolism including de novo lipogenesis, lipolysis, lipid deposition, and net lipid synthesis or net lipid loss along with lipid oxidation. Importantly, WFE-DNL-AT could be the first step to extend the homeostasis model assessment in this direction. The advantage of such modeling efforts is that the individual's entire lipid metabolism could be appropriately quantified for the clinician and connected to the processes of insulin and glucose kinetics. Further, the current modeling technique allows already for derivation of a person's individualized metrics of metabolic parameters with canonical representation [10], allowing for intra- and interindividual comparison of the parameters; observing daily changes and monitoring long-term changes of the energy metabolism with the help of a metabolic health monitoring app [11]; controlling the energy metabolism by possible implementation of a favorable adaptive control intervention with instantaneous feedback of metrics to the user to reach targeted results; and long-term analysis, cardiovascular risk assessment, and intervention planning by the healthcare team with observance of applicable clinical guidelines [2].

The ultimate outcome measure of any clinical intervention is mortality, including CVD and all-cause mortality. A CPS is empowered to calculate trends and trajectories of fat mass, which can also be translated into predicted changes of visceral fat and waist circumference. In addition to this, other important mortality predictors could be added, like heart rate variability, leading to improved prediction of all-cause mortality and sudden cardiac death [34, 35]. Measuring and tracking cardiovascular fitness in terms of exercise capacity and maximum oxygen uptake makes sense in view of the very strong inverse correlation between mortality and fitness [36] and direct correlation between insulin resistance syndrome and cardiovascular mortality [1]. Ultimately, in a fully developed CPS system, all essential metabolic variables and cardiovascular fitness measures could be tracked and used for prevention. An all-encompassing cardiovascular risk assessment by CPS could be achieved by adding traditional cardiovascular risk factors (i.e., blood pressure, smoking, age, gender, cholesterol, diagnosis of diabetes mellitus, among others), and these results would be at the fingertip of the user, who would be the owner of the data displayed on his/her smartphone. After proper consenting, secondary analysis of metabolic data could help not only clinical research but also insurance companies to calculate costs and potentially reimburse the treatment/self-treatment and improvement of risk factors for CVD. A value-based health delivery system holds potential to incentivize participants to improve their lifestyle, especially if insurance companies would honor participants with a discount on the premiums for those who were successful to lower their cardiovascular risk.

One final point is worth mentioning regarding our Lagrangian equation as it appears in Eq. (A10). The seemingly useless-looking Lagrange multiplier terms such as $\lambda \alpha w_k$, $\lambda \varrho_{Wk}$, λc_{DNLk}, and λc_{ATk} can provide important and coveted individual sensitivity values to the fixed valued constraints, i.e., sensitivity value of the energy metabolism to the parameters αw_k, ϱ_{Wk}, c_{DNLk}, and c_{ATk}. This can potentially reveal important individual sensitivity patterns to planned or executed interventions.

7. Conclusion

In this paper we presented the foundation of our mobile computing-based solutions to help better observe and control the energy metabolism. We see our approach as an appropriate response to the frustration of the public at large along with health professionals regarding perceived inadequateness of the current state of nutritional research [38] and at the same time give an answer to the call of academic authors for patient-centered individualized care [39]. Based on a rigorous examination of our methods here and in prior publications, we conclude that our mathematical modeling scheme along with the suggested computing tools is workable and appropriate to monitor, analyze, and predict individualized state variables of the metabolism, providing metrics for the quantification of the interrelationship between energy metabolism and insulin resistance which is strongly connected to obesity, fatty liver, prediabetes, metabolic syndrome, type 2 diabetes, cardiovascular morbidity, and mortality. Our mobile computing-based solutions have the potential of unlocking the vast potential of digital health. We achieved the goal of creating individualized metabolic metrics for use in mobile technology in the user's natural environment. We demonstrated that our Weight-Fat-Energy balance calculations are appropriate to predict changes of HOMA-IR. We found that the Weight-Fat-Energy balance calculations extended with the assessment of de novo lipogenesis and adaptive thermogenesis/loss predict the changes of the metabolic state variables such as Rw-ratio, weight, fat mass, and 24-h nonprotein respiratory quotient with very much acceptable accuracy. We found that the estimation of the de novo lipogenesis and adaptive thermogenesis calculations follow closely measurements that were done using a metabolic chamber for 5 days. It may just be possible to get 24-h respiratory quotient measurements in patients in their natural environment without using the metabolic chamber. We found that our analysis and predictions of the state variables of the metabolism remain valid not just at steady state but also during transitional phases. Incorporating Weight-Fat-Energy balance calculations with extended capability of de novo lipogenesis and adaptive thermogenesis assessment into mobile technologies and into a cyber-physical system [8] can provide appropriate real-time tools to monitor and optimally adjust modifiable risk factors of an individual's metabolism, allowing for planning and executing dynamic changes of behavior for optimization and control. All-encompassing cardiovascular disease risk scores can be created, tracked, modified with appropriate lifestyle changes, and used ultimately as outcome measures to improve health status. All these possibilities are applicable in resource-limited settings with minimal investment with implications for overall reduction of health costs and the potential to calculate sustainable reduction of premiums by insurers awarding compliance and efficacy in reaching and maintaining reasonable health status.

Acknowledgements

We would like to express our gratitude for the valuable related discussions with Professor Dr. John Buse, UNC-Chapel Hill. Further, we would like to thank Ilona Ori, JD, Ori Diagnostic Instruments, LLC, for editorial help.

Conflict of interest

The author declares that there is no conflict of interest regarding the publication of this paper. No specific funding was provided for this research. This research was performed as part of the author's employment with Ori Diagnostic Instruments, LLC. The author is the inventor on patent [32] and patent application [33], and the patent and patent application are owned by Ori Diagnostic Instruments, LLC.

Glossary

Measured variables

F_k	fat weight
ΔF_k	fat mass change in 24 h
W_k	weight
ΔW_k	body weight change in 24 h
EB_k	daily energy balance
ΔEB_k	positively signed energy deficit
EP_k	energy production with substrate oxidation

Derived or estimated variables

BMR_k	basal metabolic rate
c_{ATk}	adaptive thermogenesis coefficient
c_{DNLk}	fraction coefficient of the metabolized carbohydrate intake used for lipid synthesis
CI_k	carbohydrate calorie intake
DNL_k	de novo lipogenesis
DNL_k^A	de novo lipogenesis in the acute phase of energy perturbation
DNL_k^{ARC}	de novo lipogenesis in the acute phase of energy perturbation with RC diet
DNL_k^{ARF}	de novo lipogenesis in the acute phase of energy perturbation with RF diet
DNL_k^B	de novo lipogenesis in the steady-state phase of energy perturbation
$GFRC_k$	glucose equivalent of energy for needed fat (DNL) energy with RC diet
$GFRF_k$	glucose equivalent of energy for needed fat (DNL) energy with RF diet
L_k	lean mass
ΔL_k	lean mass change in 24 h
$mDNL_k$	calculated DNL using measured 24-h nonprotein respiratory quotient Rnp'_k
MEI_k	metabolically utilized energy intake
mT_k	calculated adaptive thermogenesis/thermal loss using measured 24-h nonprotein respiratory quotient Rnp'_k

PAE_k	physical activity energy expenditure via smart watch sensors
T_k	adaptive thermogenesis/thermal loss
TEE_k	total energy expenditure
R_k	R-ratio
Rw_k	Rw-ratio
Rnp_k	nonprotein respiratory quotient
Rnp'_k	measured nonprotein respiratory quotient
αw_k	first-order term coefficient of the weight-fat logarithmic relationship
λc_{ATk}	Lagrange multiplier of the adaptive thermogenesis coefficient
λc_{DNLk}	Lagrange multiplier of the fraction coefficient of the metabolized carbohydrate intake used for lipid synthesis
$\lambda \alpha w_k$	Lagrange multiplier of the first-order term coefficient of the weight-fat logarithmic relationship
$\lambda \varrho_{Wk}$	Lagrange multiplier of the energy density for weight
$\varrho_C \approx 4.2$ kcal/g	energy density for glucose
$\varrho_F \approx 9.4$ kcal/g	energy density for fat (dioleyl palmitoyl triglyceride)
$\varrho_F^* \approx \varrho_F = 9.4$ kcal/g	in the case of net fat loss; for net fat synthesis, the value is 9.4 + 2.38 kcal/g because synthesis cost of fat from glucose is added to the energy density of fat
$\varrho_{DNL} \approx 2.38$ kcal/g	energy cost for synthesis of 1 g fat from glucose
$\varrho_L \approx 1.8$ kcal/g	the energy density for lean mass
ϱ_{Wk}	energy density for weight
φ_k	fat intake fraction
χ_k	fat burning fraction

Appendix

The WFE calculation using Eq. (1) has been already introduced to the reader in [8]. WFE-DNL-AT is an extension of the WFE calculation to include de novo lipogenesis (DNL) and adaptive thermogenesis (AT). Here we are using similar annotation (see Glossary) as in [8] for the nonfat and fat balance as in Eqs. (A1) and (A2), respectively, which will include now DNL:

$$\varrho_{Wk} \cdot \Delta W_k = (1 - \varphi_k) \cdot MEI_k - DNL_k - (1 - \chi_k) \cdot TEE_k \tag{A1}$$

$$\varrho_F \cdot \Delta F_k = \varphi_k \cdot MEI_k + DNL_k - \chi_k \cdot TEE_k \tag{A2}$$

WFE calculations can provide ways to determine ϱ_{Wk}, Rw_0, and χ_k without calorie counting. In order to calculate DNL_k, the measurement of the total energy expenditure TEE_k and the metabolized energy intake and along with it the fat intake fraction φ_k are needed. In addition, knowledge of the utilized carbohydrate intake CI_k is a prerequisite as our modeling proposition for DNL follows on one hand the modeling proposition of Hall [35], and on the other hand, it follows the common clinical observation that with increasing insulin sensitivity, less DNL is generated and vice versa with decreasing insulin sensitivity (increasing insulin resistance), more DNL is generated. We use here our earlier observation that the R-ratio or the Rw-ratio is inversely correlated with HOMA-IR and mimics a measure for insulin sensitivity [8, 10, 11]. Hence the proposed model for steady-state level of DNL at baseline at day $k = 0$ is in Eq. (A3):

$$DNL_0^B = \frac{c_{DNL0} \cdot CI_0}{\varrho_{W0} \cdot Rw_0 + \varrho_F} \tag{A3}$$

Here c_{DNL0} means the fraction of the metabolized carbohydrate intake which is used for de novo lipogenesis, and its value needs to be determined. At steady state all variables must be stable over time with little or no change. The new steady state is symbolized as DNL_j^B where index $j = 1$ enumerates the steady states in sequence. Between two steady states, the DNL calculation will have acute phase component. For fasting experiments, we show here our modeling of the acute phase component. In acute phase of fasting, the production of DNL must cover the needed fat (mainly triglycerides) for the relatively unchanged fat burning rate of the body cells. The important principle for understanding the metabolic pathways is the fact that glucose is the prime ingredient for DNL production and fat cannot be directly converted to sugar. However, lipolysis can provide the needed free fatty acid and glycerol if glucose is not readily available. We found evidence by studying results of the CC study that the acute phase DNL_k^A is determined primarily by the sudden rise of the positively signed energy deficit ΔEB_k. The energy source to cover DNL_k^A is different in RF diet versus RC diet.

In case of RF diet where the required energy ΔEB_k comes from undisturbed glucose supply, the glucose is oxidized to fat during lipid synthesis to meet the demand for DNL_k^{ARF}. This is a process which would increase Rnp above 1 [31] if the ongoing fat oxidation is ignored. However, with the ongoing fat oxidation, the 24-h Rnp will not rise above 1, but it would show an increasing trend toward 1 as demonstrated in **Figure 4**. The glucose equivalent energy for needed fat energy is denoted here as $GFRC_k$ in kcal, and a simple formula in Eq. (A4) expresses at least at the beginning of the adaptation of the quantitative relationships:

$$DNL_k^{ARF} = GFRF_k = \Delta EB_k \tag{A4}$$

During the RC diet, the acute phase DNL_k^{ARC} is thought to come mainly from the fat pool through lipolysis. I modeled this in a way which clearly shows how the initial adaptation to RC diet would proportionally compare with the RF diet. This formula is in Eq. (A5):

$$DNL_k^{ARC} = GFRC_k = \frac{\varrho_C}{\varrho_C + \varrho_{DNL}} \cdot \Delta EB_k \tag{A5}$$

Here $GFRC_k$ stands for the glucose equivalent energy in kcal to meet the fat energy demand. The factor $\varrho_C/(\varrho_C + \varrho_{DNL})$ is to express that only a fraction of ΔEB_k is needed coming from fat pool with RC diet compared with glucose energy source with RF diet. The modelling equation in Eq. (A5) of the RC diet differs from Eq. (A4) of the RF diet because the Simonson's rule [13] does not apply in the RC scenario, but it will very much apply to RF scenario where lipogenesis occurs from glucose, and each gram of lipid synthetized from glucose consumes $\varrho_{DNL} = 2.32$ kcal/g energy. The graph of **Figure 5** shows well that the theoretical models in Eq. (A4) and (A5) run very close to the measured values.

Here we give our modeling of the adaptive thermogenesis which occurs with energy imbalance with metabolic changes to oppose the body composition change. In practical terms with negative energy balance, the adaptive thermogenesis T_k

diminishes the total energy expenditure; vice versa with weight gain, the adaptive thermogenesis T_k adds the total energy expenditure leading to lesser weight increase than expected by simple calorie counting. For modeling this phenomenon, we chose to use Hall's equation for adaptive thermogenesis [35] as in Eq. (A6):

$$c_{ATk} \cdot T_k = \frac{1}{\tau} \cdot \frac{MEI_{j=0} - MEI_{j=1}}{MEI_0} - c_{ATk} \cdot \frac{T_{k-1}}{\tau} + c_{ATk-1} \cdot T_{k-1} \tag{A6}$$

Here, c_{AT} means adaptive thermogenesis coefficient, a parameter which must be estimated. The value for the time constant is assumed to be $\tau = 7$ according to Hall [35]. If uncertainty exist, then parameter estimation is needed also for τ. $MEI_{j=0}$ means metabolized energy intake at baseline steady state, and $MEI_{j=1}$ is the energy intake at the end of the adaptation period when the next steady state has been reached.

The total adaptive thermogenesis sums up over time from $i = 0$ until final day $i = k$ as in Eq. (A7), assuming a quasi-stable c_{ATk} over the examined time period:

$$T_k(c_{ATk}) = \sum_{i=0}^{k} \left(\frac{1}{c_{ATk} \cdot \tau} \cdot \frac{MEI_{j=0} - MEI_{j=1}}{MEI_0} - \frac{T_{i-1}}{\tau} + T_{k-1} \right) \tag{A7}$$

For days with transition from one steady state to another, the total energy expenditure TEE_k is replaced by the sum of energy production EP_k and the adaptive thermogenesis/sink T_k as in Eq. (A8):

$$TEE_k = EP_k + T_k \tag{A8}$$

The daily energy production can be determined from Eq. (A9):

$$EP_k = PAE_k + BMR_k \tag{A9}$$

For theoretical purpose, we want to update here our proposed thermodynamic Lagrangian functional presented already in [8] to determine the unknown parameters αw_k and ϱ_{Wk} for WFE calculations. In doing so, we utilize the principles of indirect calorimetry and the principle of "least action or stationary action" [8].

The extended model WFE-DNL-AT is empowered to calculate DNL as well as AT. For this purpose the determination of the following unknown parameters are needed, αw_k, ϱ_{Wk}, c_{DNLk}, and c_{ATk}, by using known values of W_k, F_k, EP_k, MEI_k, CI_k, and φ_k. The new form of the thermodynamic Lagrangian functional contains now implicitly DNL_k as a function of c_{DNLk}, CI_k, ϱ_{Wk}, and Rw_k and the adaptive thermogenesis/thermal loss T_k as a function of the parameter c_{ATk} as shown in Eq. (A10):

$$\begin{aligned} \mathcal{L} = \sum_{k=0}^{n} & \left[\left(\varrho_{Wk} \cdot Rw_k + \varrho_F \right) \cdot \Delta F_k + MEI_k - PAE_k - BMR_k - T_k(c_{ATk}) \right]^2 \\ & + \lambda \alpha w_k \cdot \left[\Delta W_k - \alpha w_k \cdot \frac{\Delta F_k}{F_k} \right]^2 \\ & + \lambda \varrho_{Wk} \cdot \left[\varrho_{Wk} \cdot \Delta W_k - MEI_k + EP_k + T_k(c_{ATk}) - \varrho_F \cdot \Delta F_k \right]^2 \\ & + \lambda c_{DNLk} \cdot \left[\varphi_k - \frac{\varrho_{Wk} \cdot Rw_k}{\varrho_{Wk} \cdot Rw_k + \varrho_F} - \frac{c_{DNLk} \cdot CI_k}{TEE_k \cdot \left(\varrho_{Wk} \cdot Rw_k + \varrho_F \right)} \right]^2 \\ & + \lambda c_{ATk} \cdot \left[TEE_k - EP_k - T_k(c_{ATk}) \right]^2 \end{aligned} \tag{A10}$$

Note that TEE_k can be calculated from the energy balance equation which is the sum of Eqs. (A1) and (A2).

The Lagrangian functional $\mathcal{L}$ contains all energy forms entering and exiting the human body along with the subsidiary conditions for the unknown parameters αw_k, ϱ_{Wk}, c_{DNLk}, and c_{ATk} from beginning day $k = 0$ to end $k = n$ of observation. The equations with the constraints for the unknown parameters αw_k, ϱ_{Wk}, c_{ATk}, and c_{DNLk} add up to zero, and they are entered with their time-dependent Lagrange multipliers and $\lambda\alpha w_k$, $\lambda\varrho_{Wk}$, λc_{DNLk}, and λc_{ATk} as in Eq. (A10). A minimization procedure will give the estimations for the unknowns αw_k, ϱ_{Wk}, c_{DNLk}, and c_{ATk}.

Author details

Zsolt P. Ori
Ori Diagnostic Instruments, L.L.C., Durham, NC, USA

*Address all correspondence to: zsolt.ori56@gmail.com

IntechOpen

© 2019 The Author(s). Licensee IntechOpen. This chapter is distributed under the terms of the Creative Commons Attribution License (http://creativecommons.org/licenses/by/3.0), which permits unrestricted use, distribution, and reproduction in any medium, provided the original work is properly cited. CC BY

References

[1] Godsland IF, Lecamwasam K, Johnston DG. The Insulin Resistance Syndrome is an independent risk factor for CVD mortality; and an effective, clinically usable index can be derived from readily measured variables. Metabolism. 2011;**60**(10):1442-1448. DOI: 10.1016/j.metabol.2011.02.012. Epub: 02 April 2011

[2] Arena R, Guazzi M, Lianov L, Whitsel L, Berra K, Lavie CJ, et al. Healthy lifestyle interventions to combat noncommunicable disease—A novel nonhierarchical connectivity model for key stakeholders: A policy statement from the American Heart Association, European Society of Cardiology, European Association for Cardiovascular Prevention and Rehabilitation, and American College of Preventive Medicine. Mayo Clinic Proceedings. 2015;**90**(8):1082-1103

[3] Shimobayashi M, Albert V, Woelnerhanssen B, Frei IC, Weissenberger D, Meyer-Gerspach AC, et al. Insulin resistance causes inflammation in adipose tissue. The Journal of Clinical Investigation. 2018; **128**(4):1538-1550. DOI: 10.1172/JCI96139

[4] Shoelson SE, Herrero L, Naaz A. Obesity, inflammation, and insulin resistance. Gastroenterology. 2007;**132**: 2169-2180. DOI: 10.1053/j.gastro.2007.03.059

[5] Barzilay JI, Blaum C, Moore T, Xue QL, Hirsch CH, Walston JD, et al. Insulin resistance and inflammation as precursors of frailty. The cardiovascular health study. Archives of Internal Medicine. 2007;**167**(7):635-641. DOI: 10.1001/archinte.167.7.635

[6] Bonora E, Targher G, Alberiche M, Bonadonna RC, Saggiani F, Zenere MB, et al. Homeostasis model assessment closely mirrors the glucose clamp technique in the assessment of insulin sensitivity: Studies in subjects with various degrees of glucose tolerance and insulin sensitivity. Diabetes Care. 2000; **23**(1):57-63. DOI: 10.2337/diacare.23.1.57

[7] Hebert JR et al. Systematic errors in middle-aged Women's estimates of energy intake: Comparing three self-report measures to total energy expenditure from doubly labeled water. Annals of Epidemiology. 2002;**12**(8): 577-586. DOI: 10.1016/S1047-2797(01)00297-6

[8] Ori Z. Cyber-physical system for management and self-management of cardio-metabolic health. In: Type 2 Diabetes. Rijeka: IntechOpen. DOI: 10.5772/intechopen.84262

[9] Őri Z. Parametric recursive system identification and self-adaptive modeling of the human energy metabolism for adaptive control of fat weight. Medical and Biological Engineering and Computing. 2017; **55**(5):759-767. ISSN: 0140-0118; EISSN: 1741-0444. DOI: 10.1007/s11517-016-1552-3

[10] Ori Z, Ori I. Canonical representation of the human energy metabolism of lean mass, fat mass, and insulin resistance. In: 2016 IEEE 7th Annual Ubiquitous Computing, Electronics & Mobile Communication Conference (UEMCON); 2016 Oct 20–22. New York, NY: IEEE; 2016. pp. 1-8. ISBN: 1509014969, 9781509014965; EISBN: 1509014969, 9781509014965. IEEE Xplore Digital Library. Available from: https://ieeexplore.ieee.org/Xplore/home.jsp. DOI: 10.1109/UEMCON.2016.7777862

[11] Ori Z, Ori I. Fighting weight problems and insulin resistance with the metabolic health monitor app for patients in the setting of limited access

to health care in rural America. In: 2016 IEEE Global Humanitarian Technology Conference (GHTC); 2016 Oct 13-16. Seattle, WA: IEEE. ISBN: 1-5090-2433-6, 978-1-5090-2433-9; 2017 Feb 16. IEEE Xplore Digital Library. Available from: https://ieeexplore.ieee.org/Xplore/home.jsp. pp. 547-554. DOI: 10.1109/GHTC.2016.7857334

[12] Schutz Y. Concept of fat balance in human obesity revisited with particular reference to de novo lipogenesis. International Journal of Obesity and Related Disorders. 2004;**28**(S4):S3-S11. DOI: 10.1038/sj.ijo.0802852

[13] Simonson DC, DeFronzo RA. Indirect calorimetry: Methodological and interpretative problems. American Journal of Physiology—Endocrinology and Metabolism. 1990;**258**:3. ISSN: 0193-1849, 0002-9513; EISSN: 1522-1555, 2163-5773. DOI: 10.1152/ajpendo.1990.258.3.E399

[14] Müller MJ, Lagerpusch M, Enderle J, Schautz B, Heller M, Bosy-Westphal A. Beyond the body mass index: Tracking body composition in the pathogenesis of obesity and the metabolic syndrome. Obesity Reviews. 2012;**13**(2):6-13. ISSN: 1467-7881; EISSN: 1467-789X. DOI: 10.1111/j.1467-789X.2012.01033.x

[15] Őri, Zsolt: The predictability of insulin resistance and fat oxidation changes from serial measurements of weight and fat mass. Annual Scientific Meeting in Sarasota, Florida of the Hungarian Medical Association of America, October 28–November 2, 2018. Archives of the Hungarian Medical Association of America. Oct 2018;**26**(3):92. ISSN: 1070-0773

[16] Kleist B, Wahrburg U, Stehle P, Schomaker R, Greiwing A, Stoffel-Wagner B, et al. Moderate walking enhances the effects of an energy-restricted diet on fat mass loss and serum insulin in overweight and obese adults in a 12-week randomized controlled trial. The Journal of Nutrition. 2017;**147**(10):1875-1884. ISSN: 0022-3166. DOI: 10.3945/jn.117.251744

[17] Bradley U, Spence M, Courtney CH, McKinley MC, Ennis CN, McCance DR, et al. Low-fat versus low-carbohydrate weight reduction diets: Effects on weight loss, insulin resistance, and cardiovascular risk: A randomized control trial. Diabetes. 2009;**58**(12): 2741-2748. ISSN: 0012-1797; EISSN: 1939-327X. DOI: 10.2337/db09-0098

[18] Gardner CD, Trepanowski JF, Del Gobbo LC, Hauser ME, Rigdon J, Ioannidis JPA, et al. Effect of low-fat vs low-carbohydrate diet on 12-month weight loss in overweight adults and the association with genotype pattern or insulin secretion the DIETFITS randomized clinical trial. JAMA: The Journal of the American Medical Association. 2018;**319**(7):667-679. DOI: 10.1001/jama.2018.0245. ISSN: 0098-7484

[19] CCL W, Adochio RL, Leitner JW, Abeyta IM, Draznin B, Cornier M-A. Acute effects of different diet compositions on skeletal muscle insulin signaling in obese individuals during caloric restriction. Metabolism, Clinical and Experimental. 2013;**62**(4):595-603. ISSN: 0026-0495. DOI: 10.1016/j.metabol.2012.10.010

[20] Calbet JAL, Ponce-González JG, Pérez-Suárez I, de la Calle Herrero J, Holmberg H-C. A time-efficient reduction of fat mass in 4 days with exercise and caloric restriction: Fast and efficient reduction of fat mass. Scandinavian Journal of Medicine & Science in Sports. 2015;**25**(2):223-233. ISSN: 0905-7188. DOI: 10.1111/sms.12194

[21] Ebbeling CB, Leidig MM, Feldman HA, Lovesky MM, Ludwig David S. Effects of a low-glycemic load vs low-fat diet in obese young adults: A randomized trial. JAMA. 2007;**297**(19):

2092-2102. ISSN: 0098-7484; EISSN: 1538-3598. DOI: 10.1001/jama.297.19.2092

[22] Goodpaster BH, Kelley DE, Wing RR, Meier A, Thaete FL. Effects of weight loss on regional fat distribution and insulin sensitivity in obesity. Diabetes. 1999;**48**(4):839-847. ISSN: 0012-1797; EISSN: 1939-327X. DOI: 10.2337/diabetes.48.4.839

[23] Henríquez S, Monsalves-Alvarez M, Jimenez T, Barrera G, Hirsch S, de la Maza MP, et al. Effects of two training modalities on body fat and insulin resistance in postmenopausal women. Journal of Strength and Conditioning Research. 2017;**31**(11):2955-2964. DOI: 10.1519/JSC.0000000000002089. ISSN: 1064-8011

[24] Kirk E, Reeds DN, Finck BN, Mayurranjan MS, Patterson BW, Klein S. Dietary fat and carbohydrates differentially alter insulin sensitivity during caloric restriction. Gastroenterology. 2009;**136**(5):1552-1560. DOI: 10.1053/j.gastro.2009.01.048. ISSN: 0016-5085, EISSN: 1528-0012

[25] Lagerpusch M, Bosy-westphal A, Kehden, B, Peters A, Müller MJ. Effects of brief perturbations in energy balance on indices of glucose homeostasis in healthy lean men. International Journal of Obesity. 2012;**36**(8):1094-1101. ISSN: 0307-0565; EISSN: 1476-5497, DOI: 10.1038/ijo.2011.211

[26] Mason C, Foster-Schubert KE, Imayama I, Kong A, Xiao L, Bain C, et al. Dietary weight loss and exercise effects on insulin resistance in postmenopausal women. American Journal of Preventive Medicine. 2011;**41**(4):366-375. ISSN: 0749-3797; EISSN: 1873-2607. DOI: 10.1016/j.amepre.2011.06.042

[27] Thompson WG, Slezak JM. Correlations between measures of insulin sensitivity and weight loss. Diabetes Research and Clinical Practice. 2006;**74**(2):129-134. ISSN: 0168-8227; EISSN: 1872-8227. DOI: 10.1016/j.diabres.2006.03.017

[28] Weyer CR, Pratley RE, Salbe AD, Bogardus C, Ravussin E, Tataranni PA. Energy expenditure, fat oxidation, and body weight regulation: A study of metabolic adaptation to long-term weight change. The Journal of Clinical Endocrinology and Metabolism. 2000;**85**(3):1087-1094. DOI: 10.1210/jcem.85.3.6447

[29] Hall KD, Bemis T, Brychta R, Chen KY, Courville A, Crayner EJ, et al. Calorie for calorie, dietary fat restriction results in more body fat loss than carbohydrate restriction in people with obesity. Cell Metabolism. 2015;**22**:427-436. DOI: 10.1016/j.cmet.2015.07.021

[30] Guo J, Hall KD. Estimating the continuous-time dynamics of energy and fat metabolism in mice. PLoS Computational Biology. 2009;**5**:9

[31] Elia M, Livesey G. Theory and validity of indirect calorimetry during net lipid synthesis. American Journal of Clinical Nutrition. 1988;**47**(4):591-607. ISSN: 0002-9165; EISSN: 1938-3207. DOI: 10.1093/ajcn/47.4.591

[32] Ori Z. Ori Diagnostic Instruments, LLC, Assignee: An Apparatus and Method for the Analysis of the Change of Body Composition and Hydration Status and for Dynamic Indirect Individualized Measurement of Components of the Human Energy Metabolism. U.S. Patent No.: US 9,949,663 B1 Date of Patent: Apr. 24, 2018

[33] Ori Z. Ori Diagnostic Instruments, LLC, Assignee. Systems and Methods for High Frequency Impedance Spectroscopy Detection of Daily Changes of Dielectric Properties of the

Human Body to Measure Body Composition and Hydration Status. U.S. Patent Application Publication. No: US 2017/0340239 A1 Date: Nov. 30. 2017

[34] Öri Z, Monir G, Weiss J, Sayhouni X, Singer DH. Heart rate variability: Frequency domain analysis. Cardiology Clinics. 1992;**10**(3):499-537

[35] Singer DH, Öri Z. Changes in heart rate variability associated with sudden cardiac death. In: Malik M, Camm AJ, editors. Heart Rate Variability. Armonk, NY: Futura Publishing Company, Inc.; 1995. pp. 429-448

[36] Lee DC, Sui X, Artero EG, Lee IM, Church TS, McAuley PA, et al. Long-term effects of changes in cardiorespiratory fitness and body mass index on all-cause and cardiovascular disease mortality in men: The Aerobics Center Longitudinal Study. Circulation. 2011;**124**:2483-2490

[37] Matthews DR, Hosker JR, Rudenski AS, Naylor BA, Treacher DF, Turner RC. Homeostasis model assessment: Insulin resistance and fl-cell function from fasting plasma glucose and insulin concentrations in man. Diabetologia. 1985;**28**:412-419. DOI: 10.1007/BF00280883

[38] Timothy F. Kirn: Nutrition science is broken. This New Egg Study Shows Why. At turns lauded and vilified, the humble egg is an example of everything wrong with nutrition studies. UNDARK Truth, Beauty, Science. 2019. Available from: https://undark.org/2019/07/18/science-of-eggs/?utm_source=pocket-newtab

[39] Fastenau J, Kolotkin RL, Fujioka K, Alba M, Canovatchel W, Traina S. A call to action to inform patient-centred approaches to obesity management: Development of a disease-illness model. Clinical Obesity. 2019;**9**:e12309. DOI: 10.1111/cob.12309

Chapter 9

Energy Consumption Model for Green Computing

Jesus Hamilton Ortiz, Fernando Velez Varela and Bazil Taha Ahmed

Abstract

This chapter shows the environmental measurement factors that define the indicators within an architecture composed of Goods, Networks and Services (GNS). These references are obtained from the energy consumptions in the measurement processes. This lets obtaining several analyses important from the behavior in the energy consumption schemes. The application of these factors was determined from the energy generation processes, whose units are in metrics adequate for electricity and heat according to each system observed, and the same from the energetic consumption of the same data centers that helped to determine these characteristics. These indicators applied in an environment of Information and Communications Technology (ICT) define comparisons and be specified as the basis characterizing the analysis of energetic performance. The tests show the energy consumption and carbon footprint. This experiment seeks to increase the quality of services and decrease the energy consumption. This let us use efficiently the computational resources minimizing the environment impact. To achieve this target, it was used to apply indicators in green computing environment, like it can be mentioned: The Power Usage Effectiveness (PUE) and Data Center Effectiveness (DCE). With the use of these indicators can derive a generic model of energy consumption for a GNS system.

Keywords: GNS, ICT, MIB, data center, telecommunications, energy, QoS

1. Introduction

To specify a generic model of energy consumption, it is necessary to have flexibility and support in relation to the mechanisms determined in a centralized administration, by means of a network management console, which is responsible for autonomously reviewing all the individual devices and functionally they represent them [1].

To define this, it begins with the aspects that are defined to adapt and apply the standards of observation and measurement provided by the International Telecommunications Union (ITU) for a GNS environment, which is composed of Information Technology (IT) equipment, in which the behavior of energy consumption is determined, to level of the system and components [2].

In order to specification and determination of class of Management Information Base (MIB), it is necessary to know the concept of energy consumption in an architecture of a GNS system, and furthermore it is necessary to have a reference of the objects that is related to the energy consumption. In this part are applied the management parameters, which are necessary for the control of the measurement, all this functions can be included at the level of use of an interface, in accordance with the Quality of Service (QoS) parameters and management protocols required of the network [1].

The attributes that are controlled and monitored in IT team are: OFF state, SLEEP state, and ON state. If a network device is powered on, it may be transmitting or receiving packets on its interfaces. If it is inactive, it could be in a suspended state to save energy. By this, the main objective of controlling the configuration of energy management is to minimize the time in the ON state in favor of the time in SLEEP, maintaining a level of QoS in accordance with the efficient energy consumption.

Therefore, it is a good practice to have the ability of controlling the total and current mode of the ON state (active and inactive), the mode SLEEP, the OFF times, and QoS attributes defined per user, per application, and per interface. In the same way, it is necessary to be considered the ability to access the power management configuration (specifically, modes ON, SLEEP, and OFF). Also it is desirable to review all individual components of device architecture, such as network cards, processors, hard drives, and physical ports, and it is also necessary to take into account some characteristics of the traffic, like packet size and delay, that are measured in each port of the network equipment; this also affects the power consumption, and these initiatives are measured, reviewed, compared, and analyzed like as a complete system by a process of Power Management (PM) [3].

On the other hand, the task is to apply the extrapolation of the energy management model, in which a transition model of energy states is determined for everything related to the individual level of the network device, and it is clear that this can be used for a single network device, but it can also be used for an entire network, and this is achieved using a MIB for energy consumption management [1].

The mathematical process begins with the use of specific considerations of the operation of a device, among which are assuming a state of energy of these. In this case, a network device may show that there is dependence on the type of traffic that may be arriving; with this it can be determined how many subcomponents can be affected within their operational state, from this the transition time required to obtain the final state can be determined, and therefore, the optimal power control parameters can be specified, to provide the necessary QoS. Therefore, a state of **Si** can be given as a tuple [4]:

$$S_i = \left(A_J, \delta_t, D_C, Q_{i,C}\right) \quad (1)$$

This represents that system S can have different numbers of states i and have c subcomponents (e.g., devices, D); δ_t represents the transition time from one state to another state; and A_J shows the traffic characteristics, where J shows the number of packets. $Q_{i,c}$ are the optimal power control parameters of subcomponents c in the state of S_i while providing the required QoS level [4].

2. Efficient and generic energy network efficient architecture

Figure 1 functionally shows efficient and generic network architecture for model of energetic consumption. It is likely that, due to heterogeneity, in services, parameters, process results and even with the network configuration, it may be necessary to treat each of these differently in terms of energy consumption and QoS.

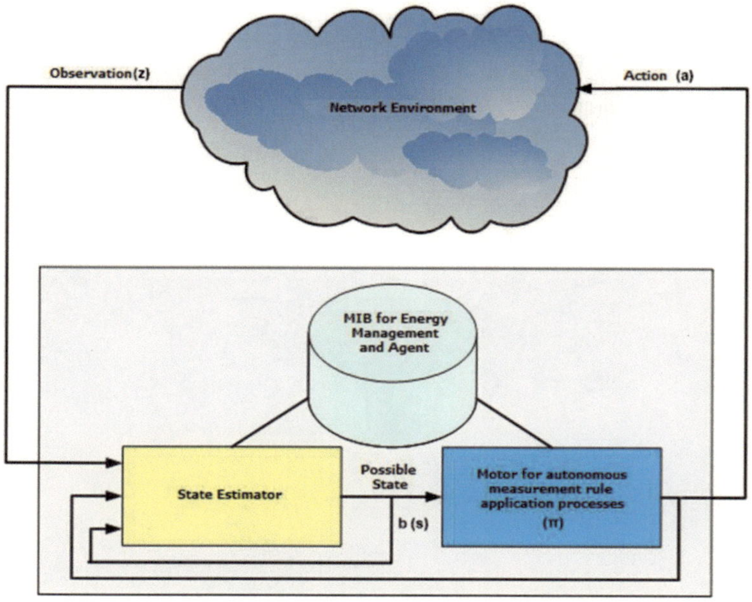

Figure 1.
Generic and autonomous architecture for energy management and QoS.

This type of model describes sequential decision tasks under conditions of uncertainty, due to that the observable system exhibits stochastic behavior. In the case of a power control policy, which calculates an action after each observation, a structural set of measures can be established within a certain period, with which the expected utility, which is maximization, can be considered. With the determination of this, an energy policy can be derived, because it is possible to calculate actions in each time step after extracting information from the respective use of the Management Information Base (MIB).

The application of the Generic Energy Consumption Model starts with the documentation of a case. If the standards are applied, can be created a structure for energy saving that can represent a good percentage of the cost reference, using the MIB model for Power Management. Like an example, the computers and printers in the resting state consume a lot of energy due to the necessary heating that they must periodically have so that they must reach again the ON state. The issue is to use prudential time and parameters so that they remain in a suspended state longer and can still operate satisfactorily when required (see **Figure 2**).

It is expected that the GNSs and production model systems will be observed, incorporating equipment in a critical way to support the operations in IT environment. It is said that criticality is related to the responsibility in the delivery of data, like a function of routers, switches, servers, etc. On the other hand, it is found that there is a network of critical equipment, which has an important role in IT operations, because these are the devices responsible for refrigeration and electrical power supply. These are referenced as generally as Non-Critical Physical Infrastructure (NCPI). In **Figure 3**, the block diagram of the observed and supervised system is presented, composed by an air conditioning unit, which is responsible for refrigeration, and the UPS's, which are coupled with an emergency battery bank.

IT devices are divided into two classes, IT1 that corresponds to telecommunications equipment, such as routers and switches, which are responsible for data transport; and the IT2 unit that corresponds to the data processing equipment and high level services such as servers.

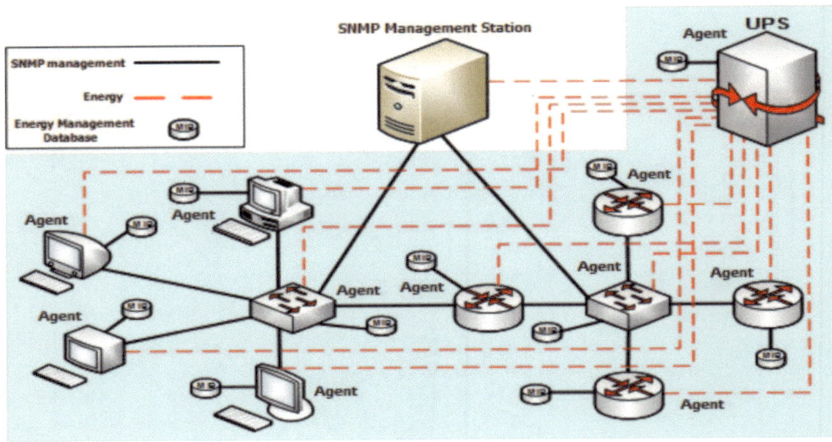

Figure 2.
Generic network scheme for energy management.

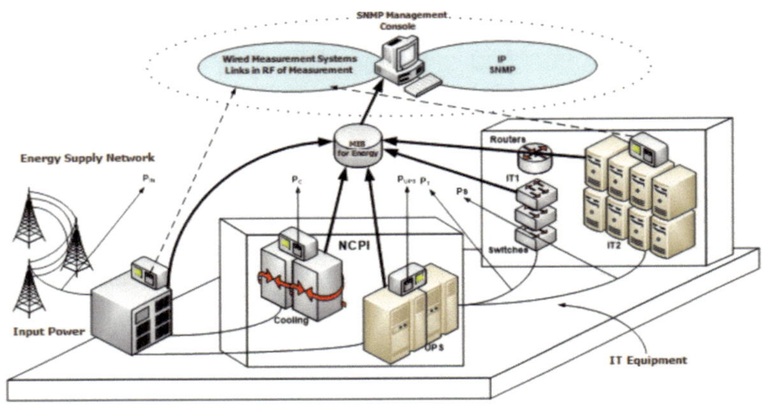

Figure 3.
Block diagram of the observed and supervised system of Goods, Networks and Services (GNS).

The total power consumption of the data center is related to the associated power consumed by each unit. The available power of the electrical network (P_{IN} of **Figure 3**) is a 110 Volt two-phase power installation. The applied power is derived by trajectories, one in series toward the power unit (UPS) and another in parallel toward the cooling infrastructure (Pc). The parallel path feeds the cooling system (air conditioning) which is important for the protection against the heat of a GNS system. The UPS systems that reach the serial power path protect the IT system from failures of supply services, provide the appropriate transition to the emergency generator system, power the IT devices, and provide the necessary energy for data processing and its transport. The power consumption of the telecommunications equipment is represented as PT and the power consumption of the servers as P_S in **Figure 3**. The energy efficiency of the data center can be broadly defined as the number of useful calculations divided by the total energy used during a process [3, 5, 6]. To consider energy efficiency, there are two types of indicators, one of them describes the efficiency of NCPI equipment, and the other type defines about useful work related to energy consumption.

The parameter PUE (Power Usage Effectiveness) is defined as the ratio between the total input powers in the facility over the power delivered to IT in the following way:

$$\text{PUE} = \frac{\text{Total Facility Power}}{\text{IT Equipment Power}} = \frac{P_{IN}}{P_{IT}}, \text{ for } \mathbf{1} < \text{PUE} < \infty \quad (2)$$

The metrics of the above equation characterize the performance or power expended in the non-critical components of the data center [3, 4]. A PUE = 2 can be interpreted that for every 1 kWh (kilowatt hour) consumed by the servers, it is necessary to enter 2 kWh (kilowatt hour) to the data center, a 1 kWh (kilowatt hour) is used in the air conditioning supply processes, uninterrupted electric power, auxiliary services such as lighting, security systems, and as a result this generates heating of power plants, ventilation systems and other minor systems.

In the current developments of virtualization systems, it is observed that the IT loads are reduced, leading to having an increasingly higher PUE. If PUE = 1, is the ideal because it indicates that for every 1 kWh (kilowatt hour) that enters the data center, the IT equipment uses 1 kWh (kilowatt hour), it is the best situation regarding the operation of the equipment of an IT system. The closer is the PUE to 1, the more efficient NCPI equipment, which are the IT elements that operate and are classified as a GNS [3].

The metric that models the efficiency of telecommunications equipment (M_T and M_T, IT) is presented in the following equation [2, 7]:

$$M_T = \frac{\sum_{i=1}^{k} b_i}{E_{IN}} \text{ [Mbits/KWh]} \quad (3)$$

In the equation, k is the number of routers in the GNS and bi is the total number of bits that leave the same router during the evaluation window [2, 7]. E_{IN} is the total energy consumed by the GNS system during the evaluation window. The evaluation window should be defined in such a way as to allow the capture of the variations in the behavior of the system observed over time. The M_T metric can measure the underutilization of routers or redundant components in the system.

The efficiency of a server is modeled as a function of average CPU utilization [2, 7]. The CPU utilization for each server in the data center is averaged in the T-time evaluation window. The metric used models the Mean Server Productivity (MSP) in relation to the total energy consumption of the system Observed GNS [5, 6], and this is conjugated in the following equation:

$$\text{MSP} = \frac{T.\sum_{\mathbf{1}}^{n} \left[U_i \cdot S_i \cdot \left(\frac{CC_i}{CB_i} \right) \right]}{E_{IN}}, \text{ [ssj\_ops/KWh]} \quad (4)$$

In the above formula, n is the number of servers and U_i is the average of CPU utilization on the server evaluation window T_i; and if the power ratio ssj_ops/sec is the 100% utilization per server in the i-th server, CC_i is the nominal clock speed of the server i CPU, CB_i is the CPU clock speed that is used for establish B_i, which is the result of benchmarking the server rate i [2, 7]. The S_i parameter (ssj_ops/sec) describes the operations of the server in one application per second and is included in the list of server specifications. With the use of this metric, it can model the productivity of the GNS systems and the correlation of the actual useful work to the maximum possible work if all the servers ran at 100% utilization.

For example, it is possible to consider these parameters like a reference of eco-efficiency, because these take into account the relationship between the service provided in terms of transmitted bits and the total energy used and consumed by the GNS system represented in joules, affecting the environment with this action. The reference factors taken into account are the amounts of information (which in this case are data and voice) derived from fixed and mobile networks, the consumption of energy for industrial purposes (control of transmission and climate), and other applications (such as electricity for office use, air conditioning and heating). The objective is extrapolating this model and applying to the case of an ICT environment composed GNS systems [8–10].

3. Energy efficiency and target values of corresponding metrics for a data center

For the selection of best practices, and to improve the current energy efficiency ratio should be defined and associated objective values that must be established on the basis actualized in terms of IT technology and infrastructure [11, 12].

The PUE (Power Usage Effectiveness), already defined, is the recommended metric for the characterization and referencing of the performance efficiency of a data center infrastructure [13–15]. It is recommended to consider the annual energy consumption (kWh) for all types of energy as the unit of measure for the calculation of the PUE. However, a category of entry-level measurement has been included in the recommendations so that operators who do not have the ability to measure consumption to use demand-based power readings and with this can be considered levels of uncertainty [16]. The measurement initiative proposes the application of this energy efficiency index for the data center.

The measurement of this parameter determines four (4) categories for a data center, which is summarized in **Table 1**.

PUE Category 0: This is a calculation based on the demand that represents the maximum load during a measurement period of 12 months. The power is represented by the reading of the demand (KW) of the output of the UPS system (or by the sum of the outputs, if more than one UPS system is installed), measured in the computer equipment during peak usage. The total power of the data center is taken by the meters and is presented as the KW demand in the utility bill. As it is an instantaneous measure, the true impact of IT fluctuations or mechanical loads can be lost. However, consistent measurement can still provide valuable data that can help in the management of energy efficiency [17–19]. The PUE category 0 can only be used for all electrical data centers, that is, it cannot be used for data centers that also use other types of energy (e.g., natural gas, cold water district, etc.) [5, 6, 16].

PUE Category 1: This is a calculation based on consumption. The IT load is represented by a total consumption reading in kWh made during 12 months at the output of the UPS system (or in the sum of the outputs, if more than one UPS system is installed). This is a cumulative measure and requires the use of

Scheme	PUE Category 0*	PUE Category 1	PUE Category 2	PUE Category 3
Location of energy measurement in IT	UPS output	UPS output	PDU output	Server input
Definition of IT energy	Electric demand IT peak	Annual energy in IT	Annual energy in IT	Annual energy in IT
Definition of total energy	IT electric demand peak	Total annual energy	Total annual energy	Total annual energy

**For the PUE Category 0, the measurements are referenced in the electrical demand (KW).*

Table 1.
Recommended categories of PUE measurement.

consumption meters in kWh at all measurement points. The total energy is usually obtained from the service company's bills by adding the 12 consecutive monthly readings in kWh, as well as the annual natural gas or other fuel consumption (converted to kWh). This measurement method captures the impact of IT fluctuations and cooling loads and thus provides a more accurate picture of the overall performance of a PUE system of category 0 [16].

PUE Category 2: This is a calculation based on consumption. The IT load is represented in kWh by a measurement by a reading during 12 months taken at the output of a Power Distribution Unit (PDU) that supports the IT loads (or sum of the outputs if more than one PDU has been installed). This is a cumulative measure and requires the use of consumption counters in kWh at all measurement points. The total energy is determined in the same way as Category 1. This measurement method provides additional accuracy of the load reading of IT by eliminating the effects of losses associated with PDU transformers and static switches [16].

PUE Category 3: This is a calculation based on consumption. The IT load is represented by a reading for 12 months and is defined in total kWh taken at the point of connection of the IT devices in the electrical system. This is a cumulative measure and requires the use of kWh consumption meters at all measurement points. The total energy is determined in the same way as category 1. This measurement method provides the highest level of accuracy for the measurement of the IT load reading by eliminating all the effects of the losses associated with the components of electrical and IT distribution that are not related, for example, the fans mounted in the racks, etc., which generates uncertainty [16].

For the normalization process, it is necessary to weight the types of energy based on the energy source. A data center that not only feeds on electricity, should be weighted according to the source of energy. This should represent the total amount of fuel that is required to operate. This should incorporate all the processes of transmission, delivery, and production losses, which allows a complete evaluation of the energy efficiency of a building. It is considered that the weighting factor for the types of energy such as electricity that is 1.0, and for natural gas is 0.35, and for fossil fuels is 0.35 normalized with electricity. Factor each component of the calculation for the PUE needs to be multiplied by the appropriate weighting factor. The weighting is obtained with the following mathematical relationship [20, 21]:

$$\begin{aligned}&\text{Weighted energy for each type of energy}\\&\quad=(\text{annual energy use}\times\text{energy source weighting factor})\end{aligned}\tag{5}$$

Because most data centers operate with 100% electric power, the recommended initiative [20] specifies the origin of the energy factors that are weighted with respect to electricity [20, 21]. This indicates that factors are developed to define the origin for each fuel and that are expressing the level of relation with the source factor of the electrical energy. This should indicate that it is agreed that purchases of electricity multiply by a factor of one, and purchases of other fuels are multiplied by the respective factors before being added to the total, and these factors are indicated like references [21].

It is required that all types of energy must be converted, by multiplying the weighting factors, and these must match in the same units before they are added. For that matter, if electricity is in kWh and natural gas is in KBTU, both must be converted to a common unit. Finally, all energy sources for all fuels must be added together [21].

The application of this to a data center has other relationships to measure the efficiency of energy consumption, as shown below:

Data Center infrastructure Effectiveness (DCiE):

$$\begin{aligned} \mathrm{DCiE} &= \frac{\mathbf{1}}{\mathrm{PUE}} = \frac{\text{IT Equipment Power}}{\text{Total Facility Power}} \\ &= \frac{\text{Annual ITKWh}}{\text{Total annual KWh}}, \text{ It is the reciprocal of the PUE} \end{aligned} \quad (6)$$

This parameter varies like this,

$$0 \leq \mathrm{DCiE} \leq \mathbf{1} \quad (7)$$

The best situation is DCiE = 1, and a DCiE = 0 is the worst situation.

The measurements of PUE, and its inverse (the DCiE), determine a reference in the input of energy to IT equipment. If an ideal PUE of 1 is considered, where the total energy that enters the data center is used by the IT equipment, and at the same instance, this quipment is not performing any computer process, can be affirmed that from PUE metric, this is an ideal situation, but in relation to the data center there is no production; for these cases, it is necessary to use other metrics, and the way the IT equipment uses energy should be reviewed.

The frequency of data collection of the PUE variables is determined from the energy consumption of a data center, and this is calculated according to the average of the efficiency of this in the same period. The measurement of the PUE is the average efficiency during a certain period. The operation cycles of the equipment in the data center are about 10 minutes, so the shortest period that gives useful information is 60 minutes. The monthly, weekly, and daily PUE measurement averages are used to determine different operating conditions of a data center, and an annual measurement of the PUE allows to easily relate the energy consumption during the useful life of this. The instantaneous measurements of the PUE must be evaluated in detail, since they are reflecting the specific situation of the operation of the IT equipment and the infrastructure equipment, in order to obtain in best form the efficiency parameters of a data center, and it is necessary to use periods of longer time [5, 6, 22–24].

To deal with this case, it is necessary to start by relating the total energy that enters the data center with respect to the energy used by the IT equipment. The following equation defines the average power [25, 26]:

$$\text{Averagepower} = \frac{\mathbf{1}}{\mathrm{N}} \sum_{\mathbf{1}}^{\mathrm{N}} \text{Power(i)} \quad (8)$$

The PUE is the point that relates the average power of the data center and the average power of it at any given time, and this same concept relates the average input power to the data center with respect to the average power of the IT equipment [25, 26]. The following equation expresses how to relate the expressed:

$$\mathrm{PUE} = \frac{\text{Average of the Total Power of Entry to the Data Center}}{\text{Average total input power to IT equipment}} \quad (9)$$

The standard deviation is determined by the following equation:

$$\sigma = \sqrt[2]{\frac{\mathbf{1}}{\mathrm{N}} \sum\nolimits_{\mathbf{1}}^{\mathrm{N}} (\text{Power(i)} - \text{Average Power})^{\mathbf{2}}} \quad (10)$$

As has been mentioned, the value of the PUE is determinate in ranges of infinity to one, the latter indicates to have an efficiency of 100% [25, 26]. If you get information from the behavior of a data center, this should tell what kind of behavior shows this system GNS, and if you have behaviors very close to a PUE = 3 or greater, these results are inefficient. The organization Green Grid establishes the allocation to each value of PUE and DCiE and relates the level of efficiency [27].

If it is necessary to make comparisons of the PUE of different data centers, it is essential to have enough information that allows to make an accurate and valid comparability. This information is supposed to compose it [22, 23]:

- Definition of the PUE measurement period: that can be yearly, monthly, daily, or per hour.

- Specification of outside temperature conditions, land area where it is located (temperate, tropical, and poles). Altitude above sea level. The climatic conditions maintain continual variations, since if it is the region of Colombia, one can speak of the model tropicalized.

- Determination of the percentage of the data center's computer load. This varies in the hours of the day, as well as every week and every month.

The loads of the GNSs of an IT system vary during a day, changing the value of the efficiency according to the time that is measured. On weekends and holidays, IT charges vary with respect to functional days, in which the efficiency varies according to the day on which it is observed. If the point of operation is determined, relating the energy efficiency according to the IT load, a data center would allow to evaluate if it is oversized in its design or has a design adjusted to the load of IT equipment.

A data center would have affected its operation, and therefore its energy efficiency is due to the combined effect of temperature, external humidity, and the computational load. The variations of the external temperature combined with the variation of the computational load determine the performance and the energy efficiency [23, 24]. This specific modeling helps to measure the electrical efficiency for a data center, where the demand vs. the losses of the energy efficiency are related, and shows how the total energy consumption is determined by the atmospheric conditions exteriors (temperature and humidity) and the load of IT

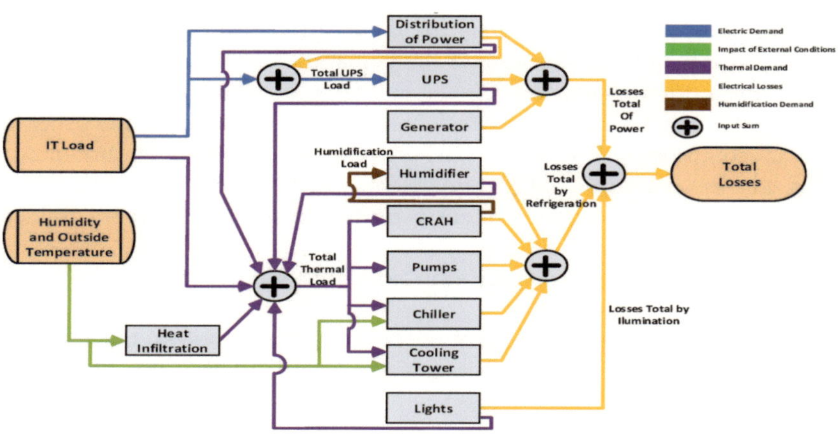

Figure 4.
Electrical efficiency measurement model for a data center. Demand vs. energy efficiency losses.

equipment; this model was obtained with SDL (Specification Description Language), see **Figure 4**.

This model is like the functional noise model in telecommunications, and in this, it determinate that total electrical losses are due to the sum of the electrical loss due to power, cooling, and lighting. The losses due to the electricity supply are the sum of the loss by the distribution, by UPS, and the other systems. The cooling system losses are in the humidifiers, Computer Room Air Handler (CRAH), pumps, ice water plant, and cooling tower. The computational charge directly impacts the power distribution and UPS system, as well as the thermal load (heat dissipated by IT equipment). External temperature and relative humidity directly impact with flows heat to the cooling systems [5, 6].

4. CUE (carbon usage effectiveness)

To estimate the energy consumption of the data centers and thus to determine the carbon footprint, the CUE is considered, which allows to determine its effect on the environment. In order to know the effect or impact that the environment generates for the energy consumption of the data centers, it must first determine this and then convert such information into carbon emissions. The carbon footprint of a data center is directly proportional to the power consumption. In the definition of this parameter, there are three variables that are considered and give a greater impact on the energy consumption, these are:

- Location, geographical location, and altitude above sea level.
- The IT equipment load (computer load).
- The energy efficiency of the infrastructure and IT equipment.

In the case of the measurement of this effect, this ratio converted to metric evaluates the emissions of CO_2 (GHG greenhouse gas) associated with energy losses in a data center. The technology with which the electricity is generated using a data center defines the emissions of CO_2 produced per kWh consumed and its expression is the term $KgCO_2/kWh$. If a certain consumption environment requires various forms of energy generation to meet the demand (as a hydroelectric, wind, nuclear, and thermal generation that consumes fossil fuels such as coal and natural gas), then that generation it considers a composition of energetic factors more extensive and likewise varies according to the state of the demand.

In peak consumption schedules, the result of the generation by energy composition can be very polluting, because this peak demand must be covered by the generation of energies that compromise fossil fuels. Carbon footprint quantifies the amount of emissions, expressed in tones of CO_2 equivalent, which are released into the atmosphere because of the development of any activity. Emissions (GHG) are divided into:

- **Direct emissions:** those that are emitted from sources that are controlled.
- **Indirect emissions:** those that are the result of the activities, but that are emitted from sources that are not controlled.

The emissions typically come from the following source categories:

- **Fixed combustion:** combustion of fuels in stationary or fixed equipment.

- **Mobile combustion:** combustion of fuels in means of transport.
- **Process emissions:** physical or chemical process emissions.
- **Fugitive emissions:** intentional and unintentional releases.

5. Measuring the green energy coefficient (GEC)

The green energy coefficient (GEC) is a metric that defines the percentage of energy that is green. To define this, there is complexity, because there are differences in the part of determination as to what is considered as the different types of renewable/green energies in reference to the region in question. This seeks to recognize as green energy anyone who is used in a data center that has a legal right defined in environmental attributes for renewable energy generation. Such legal rights are regionally recognized as certificates of green energy, renewable energy, and other similar products, which are issued by several recognized authorities who deliver the equivalent of a certificate of energy green [20, 28, 29]. Given this definition of green energy, GEC is calculated as follows:

$$\mathrm{GEC} = \frac{\text{Green Energy used for Data Center}}{\text{Total Power Source of Data Center}} \tag{11}$$

The GEC is defined by determining a maximum value of 1.0, which indicates 100% of the total energy used by the data center, and this defines the green energy [29]. Please note that for the purposes of this calculation, the total energy source consumed in the data center is identical to the PUE numerator. As with the PUE, all the energy will be notified using the same units, and the recommended unit of measurement will be kWh. The standard thermal conversion factors are used to convert to kWh are (for example, 1 kWh = 3,412 KBTU; 1 GJ = 278 kWh). Finally, because the definition of green energy is based on the legal ownership of the rights of environmental benefits, it is important to clarify that the location of the energy source does not change the calculation of GEC. For example, a data center can have a solar panel on the roof to generate power; if you sell the green energy certificates associated with this power, then you cannot be covered by the concept that electricity is green. It should be noted that this standard is measurable if it has a green power supply scheme and that helps cogeneration. In the case of the local development environment, these models are not counted.

The data center can be certified by the correct use of power, and these green energy certificates only can be achieved with a consumption of energy with a GEC = 1.00. For example, if it was considered an energy power of 130 GWh that were generated by renewable sources and this is applied to a data center, it could sell 30 GWh worth of green energy certificates without affecting the GEC of 1.00, the main idea is to have self-sufficiency, and to have an environment of electrical consumption that reverts in some way to the primary energy system, and it is to be expected that this is a considerable proportion of electrical energy, so that this leads to achieving a green certification for the management of consumption energetic [22].

6. Conclusions

- This chapter shows a model to measure energy consumption and therefore focuses on improving the energy efficiency of data centers that support the execution of services. To achieve our target, it has been used eco-Indicators in a green computing environment and it has achieved minimizing the environment impact in high percentage, with this initiative.

- The PUE/DCE are industry standards, the determination of the energy efficiency of a data center will allow comparing the degree of efficiency of an observed objective installation with other data centers around the world. It also helps to establish a point of reference that allows to control, inform, and improve continuously, in other words, to make efficient management of the resource.

- It is concluded that it is very important to develop some techniques to be used as an alternative to cloud computing in an ecological way. The following is to start the search for active migration techniques that are determined at the threshold of the infrastructures to be controlled, taking into account that it is likely that a number of necessary migrations will have to be controlled and therefore this will determine an energy consumption, that sometimes it will be expensive to use, which may affect the QoS. This involves using the help of an integrated middleware with Green Cloud Architecture, and this helps control energy consumption, but again this intrinsically adds the additional cost of middleware in the cloud computing architecture.

Author details

Jesus Hamilton Ortiz[1]*, Fernando Velez Varela[2] and Bazil Taha Ahmed[3]

1 CloseMobile R&D, Spain

2 Universidad Santiago de Cali, Colombia

3 Universidad Autonoma de Madrid, Madrid, Spain

*Address all correspondence to: jesushamilton.ortiz@gmail.com

IntechOpen

© 2020 The Author(s). Licensee IntechOpen. This chapter is distributed under the terms of the Creative Commons Attribution License (http://creativecommons.org/licenses/by/3.0), which permits unrestricted use, distribution, and reproduction in any medium, provided the original work is properly cited.

References

[1] Arciniegas H, José L, Velez Varela F. Architectures of Network Management and Telecommunications Services. University Libre Cali; 2008

[2] Jain V, Parr GP, Bustard DW, Morrow PJ. Deriving to generic energy consumption model for network enabled devices. In: India-UK Advanced Technology Center of Excellence in Next Generation Networks, Systems and Services School of Computing and Information Engineering. Coleraine, UK: IEEE, School of Computing and Information Engineering, University of Ulster; 2010

[3] The Green Grid. Proxy proposals for measuring Data Center productivity. White Paper # 13, GreenGrid Web Site. 2008. Available from: http://www.thegreengrid.org/

[4] Jain V, Parr GP, Bustard DW, Morrow PJ. Deriving a Generic Energy Consumption Model for Network Enabled Devices. 2010. Available from: https://www.academia.edu/32194810/Deriving_a_generic_energy_consumption_model_for_network_enabled_devices

[5] ITU-T Recommendation L1300: Best practices for green data centres; 2014

[6] ITU-T Recommendation L1301: Minimum data set and communication interface requirements for data center energy management; 2015

[7] Stavropoulos TG, Koutitas G, Vrakas D, Kontopoulos E, Vlahavas I. A smart university platform for building energy monitoring and savings. Journal of Ambient Intelligence and Smart Environments. 2016;**8**:301-323. DOI: 10.3233/AIS-160375. Available from: http://lpis.csd.auth.gr/publications/thanosJAISE2015.pdf

[8] ITU-T Recommendation L1310: Energy efficiency metrics and measurement methods for telecommunication equipment; 2014

[9] ITU-T Recommendation L1320: Energy efficiency metrics and measurement for power and cooling equipment for telecommunications and data centers; 2014

[10] ITU-T Recommendation L1320: Reference operational model and interface for improving energy efficiency of ICT network hosts; 2015

[11] Datacenter Consultores. Energy Efficiency Metrics for Data Center According to The Green Grid: PUE and DCiE; 2010

[12] Jaureguialzo E. PUE: Green grid metric for the evaluation of energy efficiency in CPD. In: Measurement Method by Means of the Power Demand. Madrid: The Green Grid; 2012

[13] Chromalox Precision Heat and Control. Determination of Energy Requirements for Air and Gas Heating; 2010

[14] Dunlap K. Auditing of Cooling Schemes to Identify Possible Cooling Problems in Data Centers; 2008

[15] Herrero Y, Cembranos F, Pascual M. Change the Glasses to Look at the World. Madrid: Ecologists in Action; 2011

[16] European Telecommunications Networks Operator's Association - ETNO Project, Energy Task Team; First Annual Report; 2005–2007. London; 2008

[17] ITU-T Recommendation L1400: Overview and general principles of methodologies for assessing the environmental impact of information and communication technologies; 2011

[18] ITU-T Recommendation L1410: Methodology for environmental impact assessment of information and communication technologies goods, networks and services; 2012

[19] ITU-T Recommendation L1420: Methodology for energy consumption and greenhouse gas emissions impact assessment of information and communication technologies in organizations; 2012

[20] Green Grid Project; Recommendations for Measuring and Reporting Overall Data Center Efficiency; Version 1 – Measuring PUE at Dedicated Data Centers; 2010

[21] Green Grid Project. PUE™: A comprehensive examination of the metric. 2012. Available from: https://datacenters.lbl.gov/sites/all/files/WP49-PUE%20A%20Comprehensive%20Examination%20of%20the%20Metric_v6.pdf

[22] Green Grid Project; Harmonizing Global Metrics for Data Center Energy Efficiency, Global Taskforce Reaches Agreement Regarding Data Center Productivity; 2011

[23] Rasmussen N. Implementation of data centers with high energy efficiency. Internal report No. 114 APC; 2006

[24] Rasmussen N. Imputation of Energy Costs and Carbon Emissions from a Data Center to Users of Computer Services; 2012

[25] Rasmussen N. Measurement of Electrical Efficiency for Data Centers; 2012

[26] Dayarathna M, Wen Y, Fan R. Data center energy consumption modeling: A survey. IEEE Communication Surveys and Tutorials. 2016;**18**:1. DOI: 10.1109/COMST.2015.2481183. Available from: https://ieeexplore.ieee.org/stamp/stamp.jsp?arnumber=7279063

[27] Tipley R. PUE Scalability Metric and Statistical Analyzes; 2009

[28] ISO 14040 series of standards. Standards and Technical Reports for Environmental Management - Life Cycle Analysis

[29] Green Grid Project; Harmonizing Global Metrics for Data Center Energy Efficiency, Global Taskforce Reaches Agreement on Measurement Protocols for GEC, ERF, and CUE – Continues Discussion of Additional Energy Efficiency Metrics; 2012